THIS IS CHAOS

THIS IS CHAOS

Embracing the Future of Magic

PETER J CARROLL, EDITOR

FOREWORD BY RONALD HUTTON

ILLUSTRATED BY HAGEN VON TULIEN

WEISER BOOKS

This edition first published in 2025 by Weiser Books, an imprint of
Red Wheel/Weiser, LLC
With offices at:
65 Parker Street, Suite 7
Newburyport, MA 01950
www.redwheelweiser.com

ISBN: 978-1-57863-873-4

Library of Congress Cataloging-in-Publication Data available upon request.

Cover design by Sky Peck Design
Cover art © Hagen von Tulien
Illustrated by Hagen von Tulien
Interior by Happenstance Type-O-Rama
Typeset in Adobe Caslon Pro, Begum Sans, Futura Std, Weiss Std

Figures 1–3: Wikimedia Commons, public domain; Figures 4, 7, and 10: **新纂仏像図鑑** (New Buddhist Statue Illustrated Book), published in 1931 by **仏教珍籍刊行会** (Buddhism Rare Books Publishing Association); Figures 5 and 9: MATUDA FRP with Sinobu Kurono, Figures 6, 8, and 11: Sinobu Kurono; Figures 12 and 13: Pontafon, CC BY-SA 3.0, via Wikimedia Commons; Figure 14: Illustrator Archangelus Atratus with Sinobu Kurono

Printed in the United States of America
IBI
10 9 8 7 6 5 4 3 2 1

MIX
Paper
FSC FSC® C183721

CONTENTS

Foreword, *by Professor Ronald Hutton* vii

Introduction: This Is Chaos, *by Peter J Carroll* xi

1. The Origins of Chaos Magic, *by Peter J Carroll* 1
2. Seeds of Chaos, *by Jaq D Hawkins and Peter J Carroll* 17
3. On Naive Interventionism in Magic, How Not to Do Magic, and How to Do It Well, *by Jozef Karika* 29
4. Chaos or Order: A Chaos Magic Approach to the Tarot, *by Jaq D Hawkins*. .43
5. Chemognosis Redux, *by Julian Vayne* 53
6. On the Causal Relationships Between Spirits and Archetypes, *by Jacob Sipes*. 71
7. A Path into Animist Sorcery, *by Aidan Wachter* 91
8. Thread Theory: A New Chaos Approach, *by Ivy Corvus* . . . 105
9. Collective Spirits: Egregore Entities and Online Magic, *by Dave Lee*. 131
10. Virtual Reality, Cybermagick, and the Future of Chaos, *by Lionel Snell*. 145
11. Octomantic Neuro-Hacking: A Map and a Compass for ChaoSurfing, *by Mariana Pinzón* 163
12. Chaos, Mon Amour, *by Carl Abrahamsson* 193
13. The Correspondences of Japanese Gods, *by Sinobu Kurono* . . 205
14. Chaos Witchcraft, *by Sanbre Daffowt* 221
15. The Mythogenesis of Baphomet, *by Jozef Karika* 233

About the Contributors 249

References 257

Endnotes 263

FOREWORD

BY PROFESSOR RONALD HUTTON

It is an honor to be invited to preface a fine collection such as this with a few words, and I shall begin them with two observations: that Chaos Magic is the most dynamic form of ceremonial magic in the contemporary Western world, and that its principal founder and theorist was a science teacher. Most of its character, and significance, is summed up in those simple statements. It is a kind of magic bound up intrinsically with late modernity, or postmodernity, and as such it borrows strongly from scientific methods.

In a fundamental sense, ritual magic itself is part of modernism, because it underwent a major revival from the 1890s onwards, which is still going strong. In part this revival was a reaction against the disenchantment of people's world pictures resulting from urbanization, industrialization, and rationalism. In part also, however, it was a branch of science itself, applying investigative methods to study and understand the world using esoteric lore; and it is no accident that it burgeoned alongside another new enterprise to investigate the power of the mind—that of psychology. What Chaos Magic has done is boldly update that project, by keeping pace with changes in modernity and science.

The great virtue of this collection is that its inclusion of so many able collaborators provides a consensual picture of what is distinctive and characteristic about Chaos Magic; and in this context it is especially valuable that some of them have been active in it at least since the 1990s (in the editor's case, from the beginning) and so watched it develop. It is agreed that Chaos Magic was partly a reaction against the formalism and authoritarianism of earlier traditions of ceremonial magic. In this it fitted the general revisionist mood in Britain from the 1970s onward: of examining the views of the world developed under the Victorians and hitherto dominant during the twentieth century, and deciding what was sound in them and what should be discarded.

The following features are also agreed as definitive. Chaos Magic abandons the idea of a monolithic real self and evaluates ideas according to their utility. It identifies with science rather than religion, and especially with new scientific movements such as quantum physics and chaos theory, which question traditional assumptions regarding the nature of reality and causality. It recognizes the value of working with deities (if wished) but tends to view them as creations of the human imagination. It esteems action over introspection and pragmatism over mysticism. It permits practitioners to design their own systems of belief and operation, and to change them at will. It embraces the unconstrained, the diverse, and the individual. In place of the religious concept of the faith, it puts the scientific one of the paradigm. It abolishes the opposition or division between the human mind and a spirit world. It is experimental, empirical, individual, and eclectic. It characterizes traditional views of the supernatural as useful programming languages. Above all, it places an emphasis on the power of belief to shape reality, whether or not the belief is objectively true, and on the subjectivity of human perception, and so of the world as we perceive it.

All of these characteristics can be mapped onto different temporal planes: the perennial, the modern, and the postmodern. Its experimental eclecticism and emphasis on individual choice between different belief systems is a recurrent feature of the history of magic. Both are found in the Greek Magical Papyri of the ancient world, and also in the Hermetic Order of the Golden Dawn. The alignment with science rather than religion is modernist, as is Chaos Magic's character as a working system of experimentation based on core principles, its stress on the power of the human mind, and its embrace of the concept of paradigms. The rest is distinctively postmodernist, including the doubts concerning knowledge of objective reality, and the emphasis on the subjectivity of human perception and the power of belief to shape the real. Its consistent valuation of the individual as the essential unit of society and cosmology also fits this context. No wonder it has enjoyed such success in the world that has produced both postmodernist thought and a world-changing information technology that has played up precisely those views of society and culture that Chaos Magic propagates. No wonder also that it is very much the magic of our times—and may well have the elasticity and mutability to survive as times change again.

Nothing has Absolute Truth
Anything remains Possible

INTRODUCTION: This Is Chaos

BY PETER J. CARROLL

We now inhabit an unprecedented era in which billions of individuals have access to kilowatts of power, astonishing ranges of opportunity, and vast libraries and communication facilities. The amount of available energy and information per capita in the anthroposphere began to climb exponentially in the later decades of the twentieth century. The Second World War gave a massive boost to the development of energy and information technologies, and these have remained on ever steeper upward curves ever since.

Any increase in the energy content and/or the information content of a system leads to an increase in its entropy. In the popular imagination, entropy vaguely equates to disorder, but science defines it more precisely as the number of possible microstates in a system. In this sense, entropy becomes the origin of both Order and Disorder.

"Chaos" has many meanings. It can stand as a mere cypher for Disorder. It can stand as a label for unpredictability or indeterminacy, or extreme sensitivity to initial conditions. But to an orthodox Chaos Magician, it equates to the scientific definition of entropy.

Chaos thus stands as the origin of both Order and Disorder, and the more energy and information that go into a system, the more Chaotic and creative and destructive it becomes.

So, welcome to the Pandaemonaeon of the twenty-first century, where due to energy and information overload, the Chaos in our systems, in our minds, and maybe in the entire anthroposphere tends to go exponential . . . or bust if we don't take care, but in the meantime:

> Nothing is True, and Everything is Permitted. Nothing is Sacred, and Everything is Relative and Quantum. Yet Nothing has Being, and Everything consists of Doing. So, Nothing has Absolute Truth, and Anything Remains Possible.

The information superhighways bring a cornucopia of religious, mystical, and occult ideas to the supermarkets of belief, along with a plethora of consumer choices and personal identities to select from. These all contradict each other in detail; does that make them all false, or all true, for a given value of truth?

Chaos Magic evolved as a metaphysical response to the emerging Pandaemonaeon—what shall we do, what identities shall we adopt, which beliefs shall we entertain, how shall we try to manipulate our-selfs and our environments in the Post-Truth Era?

Chaos Magicians have largely chosen to abandon self-consistency and the hung-over post-monotheist illusion of a single personal "real" self. Instead, they choose to evaluate ideas and techniques on the basis of their meaningfulness and usefulness, on the magical principle that as all ideas originate from the imagination anyway, imaginary phenomena must have very real effects.

The essays in this anthology come from a variety of orthodox Chaoists, heterodox Chaoists, and Chaos heretics. Energy and information, and hence creation and destruction, suffuse their inspirations. Taste the Chaos!

1

The Origins of Chaos Magic

BY PETER J CARROLL

Many historians have come to regard the 1970s as pivotal in the developed world. During the eighth decade of the twentieth century, so much changed in the economic, political, social, cultural, sexual, scientific, and esoteric realms. We had an unusually large number of wars, crises, coups, and economic and ideological upheavals around the world. Everything that happened in those turbulent times had its roots in the past, of course, yet rather quickly respect for authority and organized religion went into decline, quantum physics and relativistic cosmology became popular knowledge, and alternative lifestyles and cultures became almost mainstream. Plus, on top of all that, we had an explosive growth in the use of mind-altering drugs such as LSD and psilocybin, in a culture more accustomed to mere mood-altering drugs. Older certainties seemed in flux; people cast around for new ones, or tried to revitalize old ones.

As during many past epochs of cultural turmoil, a new magical revival came about, and people looked both to reinvigorate ancient

and older wisdoms and to create new esoteric insights. A publishing explosion helped the revival along, as it has helped all previous revivals along, whether by handwritten manuscript or printing press. The bookshops began to fill with reprints of material from the previous revival, which began in the latter part of the nineteenth century. The works of Eliphas Levi, the extensive Golden Dawn corpus of MacGregor Mathers, and the works of Aleister Crowley all became easily available. Plus, key texts on oriental esoterics from Tantra to Taoism hit the shelves, along with a wealth of ethnographic anthropology about animism and shamanism in remote cultures.

Chaos Magic developed out of this maelstrom, as did the modern "traditions" of witchcraft, Druidry, and various Neo-Paganisms. Chaos Magic took the radical post-modernist step of dismissing all claims to "sacred tradition" and "spiritual authority" in favor of an empirical approach. It looked at underlying theories and practices and beliefs that could give the desired results. It took the position that a magical worldview and a suite of magical techniques have developed since the earliest days of humanity, quite independently of religion, although frequently co-opted by religion or represented and expressed by religious symbolism.

Yet Chaos Magic does not seek to reduce magic to the bare study of extreme psychology and parapsychology, but rather to retain the romance of sorcery and to use the rich stores of magical thought and practical magical techniques to exploit the possibilities of extreme psychology and parapsychology more deeply.

The magical revival of the 1970s started with materials derived from the advances made following the previous revival, and it became apparent that the starting materials for the nineteenth-century magical revival all derived from the synthesis that the previous magical revival had achieved in the early centuries CE.

In those first few centuries, magicians in the Hellenic (Greco-Roman) world evolved a new metaphysic from the interface of

Judaic monotheism with neo-Platonic philosophy and classical paganism. This new Platonic–Monotheist–Pagan (PMP) paradigm spawned the magical traditions of Kabbalah, Gnosticism, and Hermetics. In this system, some sort of supreme being or essence gives rise to a series of emanations or archons or archetypes or "forms." These essences subtend a great hierarchy of subsidiary essences ranging downward from the source to include pagan god-forms, angels, demons, "elemental" forces, spirits, and the animating principles of celestial and natural phenomena, people, animals, and plants. Moreover, as thoughts and intentions also have accompanying vital essences, we can interact magically with the essences of things through the traditional practices of incantation, invocation, evocation, divination, and the casting of enchantments.

This system seems to have mainly developed in Alexandria at the mouth of the Nile, a city founded by Alexander the Great and later conquered and made even more polycultural by Rome, to become the intellectual center of the late ancient world. Here, the disenfranchised ancient Egyptian priesthood became itinerant magicians, Greek neo-Platonic philosophy flowered, and pagan religions vied with Judaic monotheism and its offshoots in various forms of Christianity. At one time Alexandria contained the largest library in the world, and it fed many of the other libraries of the classical world with its scrolls, but a poorly understood series of political and religious machinations brought about its tragic destruction. Yet, the great Platonic–Monotheist–Pagan synthesis achieved there was to dominate esoteric and magical thought in the West for nearly two thousand years.

Nothing fundamental changed in the theory and practice of magic until the advent of the Enlightenment and industrialization. The same "ancient wisdom" persisted for two millennia and merely took on local coloring. The Islamic world's esotericists used the same principles, and they helped to preserve some of the source materials during the Dark Ages in the West, but they later developed an aversion to them and began to forbid their study and practice as un-Islamic.

The Enlightenment and industrialization brought with them a highly dismissive attitude toward magical ideas and practices. From then on, magic was officially banished from a world in which science and reason could explain everything, and religion became merely tolerated as speculation about what they could not yet explain, and a useful social control mechanism.

Yet magic would not go away, and industrialism gave birth to a Romantic rebellion in which a magical revival took place. Unsurprisingly perhaps, this occurred in the heartlands of the Industrial Revolution and the centers of the new colonial empires, which housed their great libraries.

In Britain, Samuel Liddel (MacGregor) Mathers, a largely self-taught polyglot, occultist, and Freemason, assembled a huge corpus of ideas about magic drawn mainly from the vast resources of the British Museum Library. To structure his creation, he used Kabbalah as a framework and neo-Platonism as its underlying metaphysic. He managed to make this system look fairly universal by incorporating ideas and practices gleaned from Britain's imperial interaction with India, China, and Egypt. Mathers formed the Hermetic Order of the Golden Dawn to experiment with his ideas. It didn't last long as such, but it attracted a wealth of interest and talent that expanded upon the original ideas and techniques and created virtually all of the ideas that the revival of the late twentieth century started with. All subsequent Western magical neo-paganisms seem to have been built with some bricks made by, or piled up by, MacGregor Mathers.

In Mathers's great synthesis, the reality of god-forms, elemental and planetary spirits, and assorted spiritual entities and forces remains assumed for working purposes, yet operator intent and imagination also feature heavily, following the ideas of

Schopenhauer and Levi. A limited amount of operator creativity also seems implied, for example in the instructions for "designing" minor spirits for evocation. Perhaps the most seminal and significant innovation in the entire Golden Dawn corpus lay in its explicit approach to the "Assumption of God-Forms." In this, the magician attempts to personify, to identify with, and as far as possible to actually become the invoked deity through visualization, imagination, ritual performance, and altered states of mind, in order to gain some of its qualities, knowledge, and power.

This full-on approach to invocation of course has its historical precedents in activities like achieving possession by spirits and the Christian practice of the imitation of Christ, yet in the Golden Dawn system you can pick any god or goddess you like from any known pagan pantheon and give it a go. This relies on the principle that we tend to believe what we do as much as do what we believe.

I tend to regard the system that Mathers created as a precursor to Chaos Magic.

Yet, Mathers did use the then customary hidden spiritual masters gambit, and (following the lead of Helena Blavatsky) he claimed that the Order's teachings originated from Secret Chiefs, and he made appeals to the wisdom of antiquity to enhance the perceived value of what remained basically his own creative synthesis of esoteric ideas. Chaos Magic would have none of this, and its practitioners freely admitted that they had cobbled together their ideas from the best available materials and added their own inspirations, as magicians have done in every revival.

Indeed, even cursory scholarship reveals the syncretic nature of all human intellectual endeavors, whether in art, science, religion, or magic. We take what we want from the history of our endeavor, revise, reinterpret, and amend it, and try to add something more. No religion arises fully formed from divine revelation; every one of them has borrowed heavily from preceding religions. Science as we know it sits atop a vast pile of discarded and superseded theories. At least science has the good grace to acknowledge that it has frequently upgraded its ideas.

Two particular alumni of the Golden Dawn's teachings had a seminal effect on the development of Chaos Magic: Aleister Crowley and Austin Spare. Crowley took the Golden Dawn's ideas to extreme levels with increasingly wild ritual use of sex and drugs and extreme forms of meditation. This eventually led to him proclaiming himself the prophet and god of his own sociopathic religion of Thelema, replete with its own repurposed Egyptian deities and Kabbalah. After he was safely dead, Crowleyanity began to attract a substantial following in the magical revival of the late twentieth century. Chaos Magic took some inspiration from Crowley's fearless exploration of techniques for achieving alternative states of consciousness, but it rejected his theology, his philosophy of "true will," and his posthumous personality cult.

Austin Spare studied briefly under Aleister Crowley but rapidly fell out with him and his Golden Dawn–inspired magical system. Instead of looking to traditional pantheons and ceremonial and ritual procedures to express his magical inclinations, he looked to his own psychology and subconscious, his substantial artistic skills, and the romance and legends of witchcraft. Spare's work did not attract much of an esoteric following in his own lifetime. He created a wealth of striking paintings and drawings, but he wrote only briefly and enigmatically, in a strange and difficult style, about his magical ideas and practices. Yet, he had a seminal effect on the development of some magicians' ideas and practices following Kenneth Grant's publication of *The Magical Revival* in 1972, which devoted a substantial section to his work.

Spare considered that the subconscious mind held the keys to magic. He practiced divination and evocation and self-exploration by automatic drawing. He cast spells and enchantments by making abstract representations and sigils of desire, figuring that if he concentrated upon them in various altered states of mind, they would bypass conscious interference and activate the creative and parapsychological powers of the subconscious or unconscious mind. Such a mechanism can explain a great deal about how rituals and spells in general can have magical effects, and why so many spells seem like rather abstract or tangential analogies or representations of desire.

Born in modest circumstances and trying to make a living through art, Spare, unlike Crowley, practiced a lot of results magic. I found a copy of Spare's main written work, *The Book of Pleasure*, in the Library of New South Wales in Australia. I strongly suspect that Leila Waddell, one of Crowley's scarlet women, had taken it from Crowley when they split and removed it to Australia, because inside it had Crowley's Baphometic Seal, a handwritten comment a paragraph long, and an elevenfold Baphometic Cross as a signature. In the comment, Crowley dismisses Spare as a Black Magician. By this, Crowley criticizes Spare for using magic for worldly results rather than for "higher mystical purposes."

Results magic has remained central to Chaos Magic since its inception. Most of its practitioners were not born to wealth and privilege. They wanted to modify real events in their favor rather than to achieve exalted mystical states for their own sake. Thus, Chaos Magic uses enchantment to nudge the behavior of people or material reality in the desired direction, divination to yield practical information or alternative insights, evocation to bring forth servitors and entities for definite effects, and invocation to confer upon the magician some of the power, ability, and knowledge of whatever it is that they invoke.

The fifth type of magical operation, for illumination, should produce behavioral and cognitive enhancements in specifiable areas of expertise. Exactly what do you want to become enlightened about? "Spirituality" has no meaning beyond the way you lead your life.

Chaos Magic seeks to have it all three ways: to exploit the traditional forms of magical operation, to enhance them with unusual states of mind, and to explore the powers lurking within our many selfs. The most sophisticated pagan polytheist pantheons resume the whole spectrum of "gods" within us. Monotheism attempts to place one singular deity or self in charge of the totality of our universe and the worlds inside of us. Post-monotheistic cultures still largely cling to the illusion that we have a singular or "real self," despite its mutability and the internal conflicts it causes.

Religion never quite manages to separate itself from magic because people resort to religion when they fail at science or magic—if you cannot control or understand something, you can at least ascribe magical powers to it, worship it, and beg favors from it, and control other people by insisting that they do the same. Historically, this has led to the conflation of magic with religion, and for millennia magicians have tended to use the vocabulary of religion to describe their arte while those of a religious persuasion have tended to persecute any form of magic not under theological and priestly control. Religions often allowed certain exceptions for "natural magic," such as the study of the powers of stones, metals, herbs, bodily humors, alchemy, and astrology, and eventually such studies became recognized as part of natural philosophy.

Natural philosophy included what we now recognize as science and magic, with no firm distinction between the two. Gradually, a distinction did emerge: phenomena with a high degree of repeatability and an acceptable degree of causal explanation became the territory of science, while phenomena with low reproducibility and no generally acceptable causal explanation became the territory of magic. Yet, these territories lie on a spectrum, with science and magic contesting the boundaries between them. Placebo medicine, for example, can undoubtedly have strong effects, but to what extent these depend on subject belief or intent, or operator belief or intent, or all of these things remains a matter of debate. Quantum physics has likewise become fiercely disputed ground, with some magicians claiming that it reveals that the entire universe basically runs on magic and creates the illusion of material causality as a mere statistical side effect.

Psychology remains within the fold of natural philosophy as a mixture of science and magic. Psychological theories abound, with little agreement between them. More or less any explanatory scheme projected onto the mind will tend to attract some confirmation. Psychological experiments have notoriously low repeatability and exhibit a high degree of susceptibility to the intents and beliefs of operators and subjects. Predicting or manipulating people's psychological responses and behaviors remains a rather occult art with few reliable rules.

In science, "Nothing is True"—any theory that does not at least permit the possibility of being disproven does not count as science, and any confirmation of a theory simply makes it more useful until we devise something with more explanatory and predictive power. In religion, "Some Things are True"—absolutely true forever, despite all evidence to the contrary, despite that we have frequently redesigned these truths and pretended that we have not, and despite their poor explanatory and hopelessly inadequate predictive powers.

Science and magic together can give a satisfactory account of how religion works; of how belief in imaginary phenomena can have psychological benefits, facilitate social control, and lead to occasional miracles. However, religion does not tell us anything useful about science or magic at all. Thus, Chaos Magic asserts that the future of magic lies in an association with science rather than with religion.

The Chaos in the term Chaos Magic derives from the quantum insight that the universe runs on indeterminacy and randomness at a fundamental level, rather than on the basis of strict causality, as classical physics asserts, or on unknowable transcendental principles, as religions assert. In quantum terms, causality arises as an illusion created by the behavior of matter and energy in bulk. The behavior of individual quanta remains indeterminate, but large numbers render their bulk behavior reasonably predictable in the short term. Throw a single die and you might get any one of six numbers, but throw six million of them and you will nearly always get close to a million sixes. Quantum randomness can lead to a universe that seems to run on approximately causal principles most of the time.

Quantum physics has gradually seeped into the thinking end of the popular imagination since its inception in the early twentieth century. Just at the point where many felt certain that a completely mechanical cause-and-effect explanation of the universe lay in sight, along came quantum physics and upended all certainty. The further we have investigated it, the stranger it seems to become. Its most serious exponents have variously proclaimed that if you do not find it shocking then you have not really understood it, and that nobody (themselves included) really understands it. Quantum physics shows that nothing in the world has unlimited divisibility: all forms of matter and energy seem to come in minimum-sized packets that we can call quanta, and although they presumably underlie everything that we can observe in the universe, their individual behaviors seem utterly bizarre and magical.

Most of our attempts to conceptualize the behavior of quanta involve abandoning one or more of the following three ideas: reality, causality, and locality. Quanta seem not to have definite reality properties when not observed or interacting with each other. Quanta seem to do things without cause, rather acting with a randomness that appears as probability. Quanta seem capable of violating light speed and having effects instantaneously across any amount of space, defying locality. We have developed some peculiar mathematics involving so-called imaginary numbers, with some descriptive and probabilistic predictive power for the quanta. Yet, we have very little agreement about what these calculations tell us about the underlying reality of the stuff that underlies reality. Thus, a spectacular array of interpretations of quantum physics have developed, and some of these interpretations have intrigued, inspired, and delighted magicians and mystics.

Just about all of the interpretations of quantum physics (except for the basic "shut up and calculate" approach) involve some sort of "otherworld" or extra component to observable reality. The wave–particle duality provides the simplest example. It allows us to conceptualize quanta flying between events as waves of probability, but taking off and landing as discrete point particles. In this interpretation, most of reality exists in the form of a vast invisible wave-world, out of which the observable material particle world fleetingly appears to our senses and instruments. Such a vast hidden wave-world underlying reality has intriguing echoes of the astral plane or the realms of essence in esoteric and Platonic philosophies.

If we accept the ontological reality of the quantum waves that our mathematical epistemology suggests, then we open the gates to conceptual peculiarities like waves in extra dimensions of space and time and retroactive (backward in time) effects to explain the observable behavior of the quanta. Not accepting the ontological reality of quantum waves tends to lead to the even stranger many-worlds interpretation of the mathematics, which a surprising number of physicists and magicians claim to subscribe to.

Either way, the practical investigation of quantum physics has shown us that the underlying reality of the universe is far stranger and less readily comprehensible than ever previously suspected, while the theoretical, philosophical, and metaphysical investigation of it has provided a strong stimulus to esoteric and magical thinking, particularly of the Chaos Magic variety.

The formal adoption of the term Chaos Magic occurred after the discovery of Chaos Mathematics or Chaos Science. In this, we discovered that many systems evolve with extreme sensitivity to initial conditions. We now know that we cannot in principle predict long-term weather patterns without knowing the exact position and momentum of every air molecule on the planet, for starters. Dubbed the "butterfly effect," this phenomenon implies that the weather next month remains sensitive to the movements of individual butterflies today, a continent away.

Chaos Mathematics superficially appears to describe fantastically complicated systems in purely causal terms. Yet, the extreme sensitivity to initial conditions permeates right down to the quantum level, and it thus provides a ladder by which quantum indeterminacy and randomness seep upward into the world we perceive. This severely restricts our ability to predict the future because the universe itself does not know what it will do—much of the future remains indeterminate. Thus, Chaos Magic generally favors enchantment over divination, as forcing the hand of chance will usually give better results than trying to anticipate it in a Chaotic universe. Hence the principle, "Enchant Long and Divine Short."

Neuroscience, the psychedelic experience, and psychology have tended to confirm the illusory nature of the "self." We internalize the subjective cultural belief that fundamentally we consist of some sort of singular immutable entity that monitors all of our perceptions and memories, does all of our thinking and emoting, makes all of our decisions, and initiates all of our actions. However, objective analysis tells another story.

We build a self-image for good evolutionary reasons. Organisms that can form mental models of other creatures and of themselves have a big advantage over organisms that can only form mental models of their environment. Yet, such self-monitoring of various parts of the mind by other parts of the mind remains partial and very sketchy. Most of our brain remains barely aware of what most of the rest of it does, yet it adopts the convention of claiming agency for whatever thoughts, emotions, and behaviors arise, and claiming identity with them. In reality, the mind seems to consist of a raucous parliament or a pantheon of squabbling pagan gods, with no single sovereign monarch presiding over its output. This gives it huge flexibility and adaptability, but no singular "soul" or real self or true will, or any other monotheist or post-monotheist construct.

We have worlds within us; we consist of multiple impulses, agendas, and selfs. Chaos Magic seeks to explore them all by invocation. Of course, you can try to have a singular dictator self if you want one, but Chaoists tend to regard fixation with any sort of "true will," "Augoeides," or "Holy Guardian Angel" as rather limiting and obsessive, and symptomatic of possession by the demon Choronzon—your mental image of the one thing you think you should be or do, to the exclusion of all else.

Neurophysiology has elucidated the mechanisms by which the magician or the mystic achieves exalted states of extreme focus on spells and invocations or upon the object of devotion or a particular self. Extreme excitation or extreme quiescence of the mind, both brought about by essentially physiological methods, leads to the substantial inhibition of many other brain functions. Chaos Magic uses the word "gnosis" to denote the resulting state of highly focused consciousness, rather than to describe the content of any revelations or inspirations that may result from it.

One of the major themes of Chaos Magic springs from the realization that we can extend the remit of invocation from known pagan pantheons to include fantasy deities as well. This comes as a relief in an era in which it has become difficult to deny that all deities have evolved through the human imagination. Any purely imaginary beings which have at least some correspondences with human characteristics will serve just as well to extend our abilities magically. Thus, the Elder Gods of H.P. Lovecraft's Cthulhu Mythos can have their uses for the magician exploring alternative sciences. The recently realized goddess Apophenia can empower those seeking to imagine obscure, occult, or hidden but real connections between things and between ideas, so long as they remain wary of her crazy sister Pareidolia—but she has something to offer the artistically inclined.

This leads to yet another possible definition of magic—the use of imaginary phenomena to create real effects.

Eliphas Levi stressed the will and the imagination of the magician. Crowley stressed the supremacy of will. Chaos Magic asserts that will arises from the imagination. To exert so-called willpower, you simply have to imagine or absorb lots of positive feelings and thoughts about a particular course of action and imagine away most of the negatives.

In summary, the magical revival of the early centuries CE formalized a magical paradigm that we can denote as Platonic–Monotheist–Pagan, or PMP for short. This persisted until the magical revival of the late nineteenth century began to add psychological and scientific ideas to the mix. The magical revival of the late twentieth century started with this mixed paradigm and culminated in the development of what we can denote as the Quantum–Neo–Pagan (QNP) paradigm, and this underlies the theories and practices of Chaos Magic.

If a set of new ideas does not profoundly upset some people and inspire others, then it doesn't mark a significant advance. A tsunami of derision and enthusiasm always greets any real advance in magic, art, religion, or science, until the new paradigm becomes established. The ideas and practices of Chaos Magic seem to increasingly permeate the whole spectrum of the modern magical tradition.

2

Seeds of Chaos

BY JAQ D HAWKINS & PETER J CARROLL

Enough has been written about the inception of Chaos Magic during the late 1970s punk movement that the basics of the progression have become familiar, at least to those who include books and articles on Chaos Magic in their magical study. However, fully formed magical systems don't suddenly spring out of the void. The publicly known facts of the movement that grew out of English punk culture in the late 1970s were predated by a succession of innovations and societal progressions in both the US and the UK.

The seeds for the individuality and magical inventiveness that gave rise to creative chaos were planted much earlier, in the mid to late 1960s "hippy" culture, a time of social anarchy and personal exploration that played a part in shaping the coinciding magical revival of that era.

The late 1960s–early 1970s is known historically as a time of social change and experimentation, with young people pushing parameters set by previous generations with alternative belief systems and mind-altering drugs. It was also a time of sexual revolution and some bizarre fashion statements, which would carry into the 1970s punk culture. Eastern religions began to catch the imagination of Western youth, especially in the UK and US. Interest in the occult overlapped with ideas from Indian mysticism in particular, and a new magical revival gained steam.

What made this revival more significant than earlier historical magical revivals was the availability of books chronicling the practices of the occult Orders of previous eras—especially the Victorian interest in mysticism and the individualistic approaches to magic by the early twentieth-century magicians Aleister Crowley and Austin Osman Spare.

Aleister Crowley died in 1946, yet his legacy of magical experimentation continued to permeate the forming social revolution twenty years after his death, to the extent that his image appeared on the cover of the 1968 Beatles album *Sgt. Pepper's Lonely Hearts Club Band* and an article was written about him in a magazine called *International Times* (Vol. 1, Issue 33) in early 1968. This magazine, distributed by street sellers, enjoyed popularity among the hippy culture in London at the time.

In 1925, Crowley had officially been elected head of the Ordo Templi Orientis (OTO), a Thelemic group founded around the turn of the century (1895 or 1906, according to different sources). He maintained the position until his death. Crowley's penchant for drawing attention to himself and challenging the sexual mores of his time brought him notoriety but appealed to the mood of the 1960s, characterized by interest in various forms of mysticism as well as experimentation with drugs and sexual freedom.

Austin Spare was less public with his magical methods and was better known as an artist and for his unique drawing style, which played an important role in his magical practices (especially sigil magic). He privately published a few books that had some esoteric material couched within heavy social commentary, most notably *The Book of Pleasure* (1913), which detailed his process of sigil magic.

Spare died in 1956, leaving his literary legacy in the hands of Kenneth Grant, who had worked as Crowley's secretary and later as Spare's literary executor. Grant had his own ideas about magic and after Crowley's death had been involved in a dispute over the succession of the OTO, which resulted in the renaming of his branch as the Typhonian Order. It was his chapter on Austin Spare in his book *The Magical Revival* (first published in 1972 by Frederick Mueller Publishing) that brought Spare to the attention of Peter Carroll and other early Chaos Magicians.

In the 1960s, there was no internet as we know it, and common household use of computer communication wouldn't become widespread until the 1990s. However, books on magic were becoming freely available, and groups like the OTO, which numbered only in the dozens of members at the time of Crowley's death, began to spread. The eventual publication of Crowley's Thoth Tarot Deck, illustrated by Lady Freida Harris, by an American games company in 1977—complete with a contact address for the reviving OTO—may have played a seminal role in the growth of that Order.

Another magical Order, the Hermetic Order of the Golden Dawn, had effectively ceased to exist after 1908, though there would be a revival in the US in 1977. The original Order was a secret society devoted to the study and practice of Hermeticism and metaphysics, focusing on spiritual development and theurgy. One of the leaders of the Order, Samuel Liddell MacGregor Mathers, was a teacher of Crowley and a Freemason. He published several books on Kabbalah and, most notably, a tome called *The Sacred Magic of Abramelin the Mage*, which he translated from a French version of the original seventeenth-century German text in 1897.

Both the OTO and the Golden Dawn were hierarchical groups that required members to perform works to achieve higher degrees. While the Golden Dawn itself wasn't active during the 1960s revival, the writings and philosophy behind the Order were easily accessible through occult books, especially those by Arthur Edward Waite. The degree and initiatory structure heavily influenced Gerald Gardner, who effectively reinvented witchcraft (which has been around since recorded human history in various forms) and called it Wicca, going public with his new religion in 1954 when the Witchcraft Act in England was repealed.

Magic and witchcraft had a somewhat parallel development in the US, but there were notable differences. Gardnerian witchcraft was beginning to take hold in Southern California in the early 1970s through the efforts of Ed Fitch, a Gardnerian priest, but most of the country had more individualistic ideas, and the most publicized figure as the face of witchcraft during the late 1960s and early 1970s was Laurie Cabot, who was known as the "Official Witch of Salem" and wrote about her own "witchcraft" religion.

In San Francisco, a major seat of hippy and individualistic culture, two movements arose that were significant to students of magic. Anton LaVey founded his Church of Satan in 1966, which encouraged a secular and self-serving attitude toward the magical arts, and the *Principia Discordia*, the handbook for Discordianism, entered the public domain in 1965, though it was a little-known tract until much later.

Discordianism takes an irreverent attitude toward organized religion and replaces a creator god with two sisters, Aneris (Order) and Eris (Disorder). Eris is borrowed from the Greek pantheon and is the equivalent of the Roman Discordia. After the *Principia* spread in various editions with print runs in growing numbers, the "religion" of Discordianism came to public attention through a series of three novels, *The Illuminatus! Trilogy*, written by Robert Anton Wilson and Robert Shea and released in 1975.

The movement has often been equated with the rise of the Church of the SubGenius, a parody religion founded in 1970 that purports that a salesman, J.R. "Bob" Dobbs, was contacted by aliens in the 1950s and reveres him as a prophet.

During the social and magical developments of the late 1960s, Peter J. Carroll and Ray Sherwin, known for originating the Illuminates of Thanateros (IOT), were teenagers in schools in London and North Yorkshire, respectively, and not yet aware of the idiosyncratic groups forming in California. They were, however, old enough to be conscious of the social changes happening in England at the time.

Meanwhile, in 1968, a lesser-known but significant contributor to what would become Chaos Magic, Charlie Brewster, had been initiated into Thelemic Freemasonry at the ripe old age of twenty.

Charlie has a colorful and fascinating history, but most significant for the purposes of our discussion was his exposure to magical and open-minded ideas during a time of tempestuous changes in cultural attitudes. He and others involved in the Thelemic group went off to a flying saucer convention in June of that year, returning to find they had lost their apartment and the leader of the group had gone into rehab.

Subsequently, Charlie got involved in a communal group called the Tribe of the Sacred Mushroom, who practiced an experimental sort of freeform magic that he described as "of the same character as the as yet unrecognized Chaos Magic."

Between 1969 and 1973, Charlie settled down a little; he got married and had children, but lived in a quasi-communal house near Kingston-upon-Thames with other ex-Tribe members. During this period, in 1971, Pete began studying chemistry at Queen Mary's College; Ray, meanwhile, was studying English in Yorkshire and would soon travel to Egypt to teach English while pursuing his own occult interests.

Pete took an interest in hippie culture and esoterics, though he finished his degree and, after a short break, also completed teacher training. During this time, he was also attending occult lectures at a private home, where he met several people who would form a group called the Stoke Newington Sorcerers. One of the early group placed an advertisement in *Time Out* (a London-based magazine that serves as a guide to media and events happening in the city to this day) inviting people to participate in forming an esoteric group based on the Golden Dawn.

The respondents included Charlie Brewster, who had returned to London in 1974, and Lionel Snell, who had written (as Ramsey Dukes) an article about Austin Spare for a 1972 issue of *Agape Occult Review* and in 1974 published his own first book on occult philosophy, *SSOTBME: An Essay on Magic.*

It was Snell who brought a copy of Austin Spare's *Book of Pleasure* to one of these meetings, introducing several of the group members to the possibilities of sigil magic. Pete took a strong interest in experimenting with Spare's methods.

By 1976, the Stoke Newington Sorcerers had largely fizzled out, primarily due to the tenant of the apartment where they met becoming enamored with a visitor who called himself Amado Crowley (birth name Andrew Standish), who claimed to be the love child and heir of Aleister Crowley. There is no mention of him in any of Crowley's writings, and most of the group were doubtful about Amado's claims and observed that he seemed more interested in college-age young men than serious esoterics. The majority of occultists dismissed him as a fraud, and attendance began to fall away.

Pete had by this time moved into Speedwell House, a dilapidated five-story block of apartments in Deptford where his girlfriend, who would eventually become his wife, was living. The neighborhood had a history of press gangs recruiting for the navy by knocking strangers over the head; they would wake the following morning aboard a ship at sea. In Elizabethan times the playwright Christopher Marlow was famously stabbed to death in the Brown Bear pub just down the road, and the area had shed little of its notoriety since.

Charlie had moved to Sussex and visited Pete often, with the two practicing frequent rituals in Speedwell House. After a while, Charlie moved into a vacant apartment in the building and squatted it, fitting a replacement door scrounged from a derelict cinema which still bore the sign "Projection Room." Various people joined in with their esoteric activities, including amateur musicians in the area.

In the mid-1970s, the waste ground behind Speedwell House had become a campground for itinerant scrap-metal traders, who regularly had street wars with the local Deptford industrialists of the same profession. The building had been delegated to student housing (pending demolition) after the previous tenants had been evicted, but the local council had been unable to sell the land for development. It was a rough neighborhood, and Pete kept an iron bar—a retired council warden's crowbar—handy near the door. It was referred to by some of the Speedwell residents as the *Ateph Wand of Steel.* (Pete still treasures it.) Despite the violence and squalor, the creative ambience of Speedwell is remembered fondly by many who lived there during this period.

Charlie places the origins of Chaos Magic at the Deptford Olympics Goat Roast, which coincided with the Montreal Olympics in 1976. Pete had obtained a goat from the Halal Butchers in Deptford High Street for the occasion, and it was barbequed on a spit over a huge bonfire in the waste ground behind Speedwell House, followed by a live punk ensemble thrashing away and a homemade pyrotechnic display provided by some of the local anarchists.

In 1976–1977, Pete was working on the first version (the White Version) of *Liber Null*. He also wrote a few articles for *The New Equinox*, a Thelemic and Kabbalistic magazine produced by Ray Sherwin under his own imprint, Morton Press, located at his home in East Morton, Yorkshire. The book, which he talked about publishing with Ray, was intended as a magical training program and would form the basis of the practices of the IOT, the first Chaos Magic Order. The system drew from magical practices developed by the formal Orders that had gone before, as well as an animistic approach to magic common to many early cultures (often referred to as shamanism, though strictly speaking the word "shaman" is specific to certain Siberian tribes).

In 1978, the local council decided to try once again to demolish Speedwell House, and the various residents began to move on. Pete decided to go on a backpacking trip to India, then Australia, leaving the publication of *Liber Null* in the hands of Ray, who also published the first edition of his own *Book of Results* in the same year. Charlie and his family were rehoused by the council, as were many of the Speedwell residents. Pete described his motivation for overland travel to India as largely motivated by adventure and curiosity, although he did spend many months studying esoteric matters in the Tibetan Library at Dharamsala in the Himalayan foothills.

Much of the history after this point has been written about elsewhere, but it's worth noting the combination of events and influences that fed into the development of Chaos Magic. The old Orders provided plenty of material to learn formal magical ritual, but Crowley and Spare each in their own way had demonstrated that the religious overtones and strict adherence to ritual methods practiced by these groups were not a part of what makes magic work. While Discordianism was still more of an American practice, Pete had become familiar with the irreverent philosophy through reading *The Illuminatus! Trilogy*. The ingredients of a completely individualistic approach to magic were in place.

Add to that the societal changes brought about by the 1960s hippy movement and the 1970s punk culture, and the genie had been let out of the bottle. Young magicians like Peter Carroll, Charlie Brewster, and Ray Sherwin, who were well educated and independent minded, were able to transcend the superstitions and control techniques that had been practiced by priestly classes in the past. The time was ripe for a completely new attitude toward magic.

Pete returned to England in 1980 and moved to Ray's village in Yorkshire, where they produced a revised (Red Edition) version of *Liber Null*. The IOT had notionally formed with the publication of the White Edition, but only really got going after Pete's arrival.

Pete stayed in East Morton for a year, during which he wrote *Psychonaut*, which referenced catastrophe theory—a sort of precursor to chaos math, which, unknown to the young magician, was in the process of being developed at IBM. Then, in 1982, he spent another six months traveling in India before returning to settle in Bristol.

Ray had "excommunicated himself" (in his words) from the IOT, though not from Chaos Magic, due to a disagreement about the structure of the Order, and Pete evolved the name of the group to the Magical Pact of the Illuminates of Thanateros, or just "the Pact." By the mid-1980s, the Pact had Temples in England, Germany, Austria, Switzerland, and Australia. The first Temple in the US formed in 1988.

In 1986, *The New Equinox* morphed into a new Chaos Magic–focused magazine, *Chaos International*, which published articles written by Chaos Magicians from various backgrounds. Christopher Bray, who owned a shop called the Sorcerer's Apprentice in Leeds, continued to produce the Red Edition of *Liber Null* until 1987, when Pete sent copies of *Liber Null* and *Psychonaut* to Samuel

Weiser Publishing, having secured a recommendation from Israel Regardie via Gerald Suster that they consider the books. The combined edition was accepted and published in hardcover.

Following the Weiser publication, the books traveled to the German-speaking world, and Pete was contacted by Ralph Tegtmeier, who asked him to join his seminar circuit. The IOT and Chaos Magic had begun to spread.

Having remained friends with Pete and continued in his own path of Chaos Magic, Ray Sherwin organized the first Chaos Symposium in 1987, at the Griffin Hotel in Leeds. A few other Chaos Magicians put out chapbooks with their own ideas, and then, in 1988, James Gleick released *Chaos: The Making of a New Science*, bringing the elements of Chaos Science into the public arena. The correlation between this new approach to nonlinear dynamics and results-oriented methods in magic, which was already being referred to as Chaos Magic, was immediately apparent.

Computer-based communication was developing rapidly by this time, and bulletin board systems (BBSs), the precursor to internet forums, began to spring up, some of them focusing on sharing resources and conversation about magic.

Soon after the release of James Gleick's book, Jaq D Hawkins was asked to write an article on Chaos Magic for *Mezlim* magazine, having shown an interest in the subject after reading all of the works of Austin Spare, Pete Carroll, and Ray Sherwin and various other chapbooks, including an early version of Phil Hine's *Prime Chaos*.

As part of her research for the article, she wrote to Pete and to a Pagan contact magazine, asking if there were any Chaos Magicians among its subscribers. Pete asked Charlie Brewster to answer her queries, and Charlie replied with a four-page typewritten letter relating the development of Chaos Magic from his perspective. Ray responded to the query in the Pagan magazine and became a regular correspondent, providing his own views and a long letter describing the history from his own point of view.

The resulting article, "Defining Chaos," was published in *Mezlim* Vol. 2, No. 2, after which Jaq became a regular contributor to the magazine. One of Jaq's online contacts asked permission to post a copy of the article on his BBS board, but credited the authorship to Jaq's BBS handle, Mark (or Markie) Chao.

In 1990, Charlie Brewster published *Liber Cyber*, which was to be his most significant work (though he had written several articles of interest). The 1990s saw Chaos Magic in various forms spread widely due to the rapid development of internet communication. No longer exclusive to the IOT, the approach and ideas generated by the writings of Pete and Ray, and often referenced directly from Austin Spare, gave rise to personal interpretations as well as copycat practices. Other small groups of Chaos Magicians sprang up in localized areas to practice their own experimental forms of magic.

Jaq expanded "Defining Chaos" into a book, *Understanding Chaos Magic*, which was published in 1996 and cannibalized into *The Chaonomicon* in 2013, after the original publisher collapsed. This provided a simplified explanation of the basics of Chaos Magic to a widespread and rapidly growing new generation of Chaos Magicians.

In the interim and afterward, more authors published books offering their personal perspectives on Chaos Magic, often imposing their previous and sometimes very ordered magical influences onto the freeform attitudes encouraged by the original form. Several people declared Chaos Magic "dead" in the early 1990s, and two or three attempts were made to commercialize it, yet the spirit of Chaos Magic continues to his day.

Peter Carroll announced his formal retirement as head of the IOT in *Chaos International* No. 12 (1995), though he has continued to write books on magic and to teach magic through Arcanorium College, an online educational facility that began as a forum-based college, then evolved into private tutoring.

Charlie Brewster died in 2013 from heart-related problems, but left behind a wealth of fascinating material. Ray Sherwin died in July 2023 in Las Palmas, Gran Canaria. His early books, *The Book of Results* and *Theatre of Magick* (published in 1983), continue to be of great interest to Chaos Magicians, but his writings afterward shifted in focus toward his essential oils business and his poetry, and eventually to self-indulgence and conspiracy theories.

Other early Chaos Magicians, such as Dave Lee, continue to write and teach or participate in talks at large events. The publication of the magazine *Chaos International* officially ended with issue 23 in 1999, until Ian Read, who had been editor since the third issue, decided to produce more issues on an ad hoc basis. This lasted as far as issue 26, produced in the early 2000s.

Far from dead, Chaos Magic continues to thrive and evolve, having freed practitioners of magic from the constraints of ages past. There is no going back to medieval ignorance or priestly gatekeepers. That bridge has succumbed to the eternal fire.

3

On Naive Interventionism in Magic, How Not to Do Magic, and How to Do It Well

BY JOZEF KARIKA

I

Let me begin with two brief clarifications. First, in his book *Antifragile: Things That Gain from Disorder*, Nassim Taleb describes the organic–mechanical dichotomy, or the even more effective distinction between noncomplex and complex systems:

> *Artificial, man-made mechanical and engineering contraptions with simple responses are complicated, but not "complex," as they don't have interdependencies. You push a button, say, a light switch, and get an exact response, with no possible ambiguity in the consequences, even in Russia. But with complex systems, interdependencies are severe. You need to think in terms of*

> *ecology: if you remove a specific animal you disrupt a food chain: its predators will starve and its prey will grow unchecked, causing complications and series of cascading side effects. Lions are exterminated by the Canaanites, Phoenicians, Romans, and later inhabitants of Mount Lebanon, leading to the proliferation of goats who crave tree roots, contributing to the deforestation of mountain areas, consequences that were hard to see ahead of time. Likewise, if you shut down a bank in New York, it will cause ripple effects from Iceland to Mongolia. In the complex world, the notion of "cause" itself is suspect; it is either nearly impossible to detect or not really defined. (Taleb 2012, 56)*

Remember that in complex systems it is hard to see the arrow from cause to consequence, making much of conventional methods of analysis, in addition to standard logic, inapplicable. We just cannot isolate any causal relationship in a complex system.

Second, when a treatment causes more harm than benefit, we call it *iatrogenic*. Naive interventionists are zealous people who come armed with solutions to address the first-order consequences of a decision, but end up creating worse second- and subsequent-order consequences.[1] Shane Parrish writes on this:

> *Some examples are easier recognized than others. For example, when the negative effects are immediate and visible and appear to be a direct cause-effect, we can reasonably conclude that the intervention caused a negative effect. However, if the negative effects are delayed or could be explained by multiple causes, we are less likely to conclude the intervention caused them. . . . The key lesson here is that if we are to intervene, we need a solid idea of not only the benefits of our interventions but also the harm we may cause—the second and subsequent order consequences. (Parrish, n.d.a)*

Such actions usually result from one or more of the following reasons (besides malice): an inability to think through problems, separation from consequences, a bias for action, and no skin in the game. You can read more about this in Parrish's two articles (n.d.a, n.d.b), but you'd also better read *Antifragile*, which is a priceless compendium of Baphometian antifragilistic, opportunistic, ruthless, and selfish wisdom—the real bible of Baphomet.[2]

For now, it is enough to remember that intervention should only be used when the benefits visibly outweigh the negatives. A simple rule for the decision maker is that intervention needs to prove its benefits, and those benefits need to be orders of magnitude higher than what is offered by the natural path.

II

Okay, now we're set, we can get going. Both of the above—ignorance of the difference between complex and noncomplex systems, and naive interventionism—are strikingly common in the magical, occult, and esoteric scene. They are almost synonymous with it.[3] The best illustration in which these two errors achieve their glorious apex is *The Fourth Way*, P.D. Ouspensky's "lucid explanation of the practical side of G.I. Gurdjieff's teachings." These teachings go far beyond the usual degree of ignorance, which is normally ambitiously high in occult circles. The basic idea (stripped of sheer cosmic eso-insanities, cultish language, and social control) looks like this:

> *Man is a sleeping machine and therefore behaves mechanically, erroneously, and inefficiently. Thus, he needs to be constantly self-aware in order to overcome his reflexive centers because, after all, we know better how evolution should go. So, we develop exercises for which he needs to be as aware as possible and let him practice them until he's delirious. We reprogram the machine to work better. What could possibly go wrong?*

This is exemplary naive interventionism and failure to distinguish between the mechanical and the organic. "Man is a machine" is a fundamentally erroneous thesis. All devastating consequences derive from it. Man is an organic and therefore complex system; to apply a mechanistic and linear "if A, then B" approach to man is a sure recipe for disaster. The Fourth Way, and all "systems" that seek to reprogram the brain and strive for "more consciousness," go against the eons of Baphometian blind evolutionary wisdom. Sure, we can say to ourselves that from now on we will behave "more consciously." We will decide everything intentionally. Let us stretch our willpower, and we will be aware of ourselves all the time—hurrah!

What does it mean in reality? In his book *Focus: The Hidden Driver of Excellence*, Daniel Goleman details the detrimental effects that constant focus has on creativity and other aspects of life. Because of our biology and regulatory systems, in addition to a focused state, we need to spend a lot of time in a deliberately unfocused state and with an aimlessly wandering mind if we want to optimize our performance and quality of life.[4]

According to behavioral scientist Colin Camerer: "We have this layer of prefrontal cortex made just for us, sitting on top of this big animal brain. Getting this thin little layer to handle more is unrealistic. . . . It's already overtaxed" (Duke 2019, 13).

Decision strategist Annie Duke sums it up best: "Making more rational decisions isn't just a matter of willpower or consciously handling more decisions in deliberative mind. Our deliberative capacity is already maxed out. . . . The challenge is not to change the way our brains operate but to figure out how to work within the limitations of the brains we already have" (Duke 2019, 13–14).

So what about Gurdjieff, Ouspensky, and the other Fourth Way followers? Sure, they screwed up big time. Many of them, previously high achievers and creative individuals, have ended up as exhausted wrecks incapable of even the simplest creative act. Moreover, they have also lost the ability to make decisions about their own lives; their autonomy has completely disappeared. They have not attained any higher state through "The System"; no geniality, superhuman productivity, or creativity. On the contrary, they have degenerated from their original state to an inferior one. For many, falling for Gurdjieff's hook became the greatest tragedy and mistake of their lives.[5] Gary Lachman describes their harsh fates in the book *In Search of P.D. Ouspensky: The Genius in the Shadow of Gurdjieff*. It's a scary but educational read about how dangerous naive interventionism can be—especially in the area of directly messing with one's own brain.

III

Another example of naive interventionism is so-called dream yoga: i.e., continuous, long-term cultivation of lucid dreaming. Such interference with a complex phenomenon such as human sleep can produce a variety of unintended effects and hidden harms in the long term. There's nothing wrong with spontaneous lucid dreaming from time to time, but systematic cultivation of dream yoga is unnatural—an adjustment, a reprogramming of the brain's default setting, which is the result of eons of Baphometian evolution, based on one's own intention and/or some fancy theory about "improvement" or "liberation." Such tinkering with Baphomet's blind wisdom and homeostasis can bring beneficial results only rarely, with detrimental or devastating results more likely. As Nassim Taleb has noted: "It seemed to me that it was an insult to Mother Nature to override her programmed reactions unless we had a good reason to do so, backed by proper empirical testing to show that we humans can do better; the burden of evidence falls on us humans" (Taleb 2012, 338).

Few in the occult circles seem to think about it. We cannot rely on scientific studies here; I have found only one (Vallat and Ruby

2019), and it encourages great caution. We do not know at all what impact such long-term cultivation of lucid dreaming has on sleep architecture and regulatory processes of the brain,[6] especially for the social functionality of the Western practitioner. This has to be considered quite differently from the situation of the monks in Tibetan monasteries, who consider themselves already dead to the ordinary world and no longer need to function in it. This also applies to the hard-core practice of vipassana, zazen, yoga, qigong with reverse breathing, and other Eastern techniques.[7] They need to be delved into relatively lightly and gradually if ever; or stick to practices whose impact has been tested by the scientific method, at least to some extent, on a sufficient sample of Western practitioners, and over a sufficiently long time scale.[8] It won't give certainty, but at least a slight increase in safety. Avoid wild and haphazard experimentation with your own brain, the settings of which determine everything else in your life. Concentration, thinking, visualization, etc. are biological functions, just like digestion. Therefore, they should be considered in terms of intake/output, strain/regeneration, atrophy/overload relations.

IV

If we examine the real consequences of the various "systems" and their fancy theories on the lives of those who have applied them rigorously, it shows just one conclusion: the greatest chance for success (or at least not ruinous failure) lies in not delving too deeply into the default, natural bodymind adjustment.[9] If you want to step in this direction at all, work mostly on physical and mental health (in this order), psychological flexibility, openness, and creativity. Attention control and concentration are also useful, but in moderate doses.[10] Focus on the really important stuff—health, active body, nutrition, climate, natural sciences, arts and crafts, useful skills, worldly knowledge, and magic, surely not "spirituality." Yeah, you can self-improve: through exercise (physical and the very few mental exercises that work), studying the no-bullshit domains and implementing useful stuff, learning how to work with habits and how to create beneficial flow activities for yourself, and well-mastered magical techniques.

Among the practitioners I have known, no "system" has transformed or elevated anyone, but they have harmed many. The risk/gain ratio here seems to be asymmetrical in favor of harm and adverse effects. Those who have achieved something significant while practicing some "system" or tradition have not achieved it because of it. Rather, both the practice of the "system" and the achievements were expressions of their selves (plus luck). They just harnessed the uniqueness of their multi-selves, some specific combination and proportion of their chaotic intrinsic drives.[11] Both their achievements and the practice then increased the complexity of their multi-selves in a loop, but the initial impulse came from them, not from the "system."

If there is anything else helpful to this, it's directing the focus outward, away from self-absorption. Meditate through creating, sport, or other people you give your full attention to, thus cutting off the inner monologue and obsession with yourself. This also gives you some real superpowers.[12]

V

Now that we've amused ourselves with how naive practitioners of other traditions are, let's take a look at us. Chaos Magicians are in just as dire a position, if not worse. Magic is a risky venture, and Chaos Magic even more so.[13] Therefore, we need to think carefully about this matter—as they say, "measure twice, cut once." In Chaos Magic circles, I frequently encounter more of the approach "cut it all on each side like a maniac, then measure the scraps." Not so fast, young wands; the consequences can become serious.

The soundest advice: don't bother. Magic isn't worth it in most cases. Most magicians don't achieve in their lives with magic even that what most civilians achieve without magic. Much less, actually.[14] The opportunity cost of time and energy spent on magic is the sum of *everything else* you could have spent that time and energy on instead. Think about it, really. For the vast majority, you'll get further in your preferred ventures without magic than you will with magic. Or you can reach for those elements of magic that today have become part of psychology, persuasion, marketing, self-help trends, and so on. This seems a fairly low-risk path, at least relative to launching real conjurations. If you can't resist and your body pushes you permanently and irresistibly toward raw magic, then at least consider the following.[15]

Bodymind works as an interconnected complex system. That means we cannot predict what any magical intervention will directly or indirectly cause (unintended, delayed consequences and hidden harms). Moreover, in a complex system, nonlinear relations apply. "Once again, 'nonlinear' means that the response is not straightforward and not a straight line, so if you double, say, the dose, you get a lot more or a lot less than double the effect—if I throw at someone's head a ten-pound stone, it will cause more than twice the harm of a five-pound stone, more than five times the harm of a one-pound stone, etc.," writes Taleb (2012, 268).

This raises at least two important points. First, in bodymind everything seems functionally interconnected. There is no such thing as isolated change in the psychosphere. You can't think naively linearly here. For example: I need more eloquence in this period, so I will invoke Mercury repeatedly and thus achieve the desired change; problem solved. Repetitive stimulation and emphasis of the mercurial drive will trigger the response of all seven other major drives (sub-selves personified as god-forms).[16] The same rule applies to working with sigils and any personal traits. If you strengthen/stifle one too much, you change all the others, especially the opposing trait.[17] So with any magical act, especially repeated in one direction or aimed at one focal point, always assess and observe the entire ecology of your psychosphere. Use active imagery, dream analysis, or divination to do this.[18] If you perceive that some other entity in the psychosphere is resonating and demanding attention as a result of your conjurations, take care to compensate. Usually you have a few days or weeks to handle it. If you ignore it, the balance then tends to correct itself, and quite violently, often in a way that you don't find amusing.[19]

Secondly, since we act in a nonlinear domain, we cannot estimate in advance the strength of the triggered effect. This means that you can conduct five Mercury invocations in a row over a period of time, each time with a mild result, but the sixth one will trigger an effect a hundred times stronger. Or a thousand times, for that matter. This effect can go both ways, hurtful or beneficial, depending on whether you are fragile or antifragile to it (in both cases just up to a certain point).[20] Sink or swim. If you want to prepare yourself, it seems vital to take steps in the direction of antifragility, or at least robustness.

Through magic conjurations (and all flow activities) you increase your complexity, but it's a simple organism that is robust. Therefore, magicians must fortify themselves with protective means. These are not what you might expect; forget protective talismans and circles, be real. "The means by which Julius Caesar defended himself against sickliness and headaches: tremendous marches, the most frugal way of life, uninterrupted sojourn in the open air, continuous exertion—these are, in general, the universal rules of preservation and protection against the extreme vulnerability of that subtle machine, working under the highest pressure, which we call genius," wrote Nietzsche (1982, 532). This applies to the magician as well; there's no evasion unless you want to end up like most magicians who ignore it—i.e., miserable. For more practical tips, consult Taleb's *Antifragile* and Seneca's *Letters from a Stoic.*

This becomes more and more valid with increasing age. It's the same, generally, with financial or health matters—the younger you are, the more risks, errors, and harms you can endure. You recover quickly, and you will have enough time to regenerate even from serious harms. However, the older you get, the fewer mistakes you can afford. Therefore, the older you get, the more carefully you should consider when and whether to magically intervene at all. Pete Carroll (2022) summed it up this way: "I can only say, never conjure just for the sake of it or to see what will happen, think carefully before conjuring and only conjure for something you really want and can deal with."[21]

This doesn't mean that we should not magically intervene at all. But, especially the older you get, consider each such intervention more carefully. A rule of thumb that Nassim Taleb demonstrates well in the domain of medical interventions applies here:

> *Only resort to medical techniques when the health payoff is very large (say, saving a life) and visibly exceeds its potential harm, such as incontrovertibly needed surgery or lifesaving medicine (penicillin). . . . For in these cases medicine has positive asymmetries—convexity effects—and the outcome will be less likely to produce fragility. Otherwise, in situations in which the benefits of a particular medicine, procedure, or nutritional or lifestyle modification appear small—say, those aiming for comfort—we have a large potential sucker problem (hence putting us on the wrong side of convexity effects).*[22] *(Taleb 2012, 336–337)*

In other words, do not conjure for trivialities such as "minor adjustments," comfort, and other pettiness. In such cases, the iatrogenic effects induced are quite likely to cause more harm than the potential gain. The more grim and bleak the situation, the more serious the obvious harm, the more magical intervention (in addition to secular measures) becomes justified, and vice versa. Achieving something you really can't live without, something you deeply, innately desire, can also be considered critical. Spending a lifetime without it seems the biggest harm. Most often it involves some kind of experience or self-realization, but this desire must really spring from the depths, from flesh, your organic base. It should not be a shallow, petty wish—in such cases, it is not regarded as sufficiently serious.

Also, the less often you conjure, the more time you allow the bodymind to regenerate and adjust to its natural homeostasis. By putting enough time between each magical intervention you increase your chances of avoiding nonlinear effects—which is a double-edged sword. By doing so, you dodge both the extremely harmful and the extremely beneficial (potential) nonlinear effects of conjuration. It's your call.

The trouble is that if you become a "crisis magician" and thus only reach for magic in critical situations, your technique almost surely won't be thorough, synchronous, and sharp enough just when you need it most. This means that the risk of iatrogenic effects or no effect at all after the conjuration will increase significantly. Therefore, you should perform the magical ritual frequently and irregularly (more on that in a moment), but only grounding, encircling, and empowering, without conjuring.[23] Ongoing training in concentration, attention control, mindfulness, visualization, and intentionality also proves useful as an addition to the ritual.

Finally, consider this as well: "Volatility of an exposure can matter more than its average—the difference is the 'convexity bias.' If you are antifragile (i.e., convex) to a given substance, then you are better off having it randomly distributed, rather than provided steadily" (Taleb 2012, 342).

Again, this seems (anecdotally) to apply to magical action too. Randomly distributed exposure to conjurations or magical exercises seems more effective than steady, regular exposure, at least if you're antifragile to them. So, if you're aiming for nonlinear effects, randomize your conjurations in time[24] as well as in type. A novice probably needs to learn the conjurations in a linear sequence; i.e., enchantment first, then divination, evocation, invocation, and illumination. However, once learned, they seem more efficient if performed randomly. Prepare two or three important goals for each of the five types of conjurations. Then try experimenting first at a fixed time of the day with a randomly chosen conjuration. You can roll a die, assigning each number from 1 to 5 a different type of conjuration. (You can assign the number 6 to the conjuration you need to practice the most. It will then be represented by two numbers on the die, increasing the likelihood that it will be chosen. Or, a 6 can mean a toss once more, or no conjuration today.) That way you won't know in advance which conjuration you'll conduct, or how they'll be distributed over the course of a week or two. Then take a long magic rest. Once you've recovered sufficiently, you can try again, this time at different times during each day. At another occasion you can practice regularly and steadily for a while and then suddenly break the pattern; completely shatter it. Such techniques will bring volatility of exposure to your practice. Observe what effect it has.

4

Chaos or Order: A Chaos Magic Approach to the Tarot

BY JAQ D HAWKINS

Magical systems generally speak in symbols. Some of these can be very ordered, like the classical Tarot, which is a particularly ordered system that works with symbols from a variety of other systems, including Astrology, Christianity, the Kabbalah, and Hermeticism. The cards are usually used for divination among modern witches and magicians, but it wasn't always so.

There are various stories about the origin of Tarot cards, but the first verifiable decks were drawn by French and Italian artists in the later part of the fourteenth century. These were simple cards made for parlor game play and the suits, apart from the face cards (King, Queen, Knight, and Knave, also called Page or Jack), were illustrated only with their own symbols—Cups, Wands (or Batons, Rods, Clubs), Swords, and Coins (Pentacles, Disks)—rather than with the esoteric imagery that was later added.

Trump cards were introduced from the early fifteenth century, often made to order for wealthy families. Some of these were painted to resemble family members—usually barons or dukes, as it was costly to have personal cards painted and only affluent people could afford them. Some of the sets of cards still in use today resemble members of the Visconti family of Milan, who commissioned several sets of cards.

Divination with playing cards didn't start to become popular until the late sixteenth and early seventeenth centuries. The esoteric attributes we associate with Tarot today evolved over the following two centuries, moved through significant bifurcations by a small number of historical figures.

The first of these was Antoine Court de Gebelin, a French Freemason who published an analysis in 1781 in which he claimed the symbols in the Tarot were derived from esoteric secrets of the Egyptian priests. Evidence for this assertion is sadly lacking, but it instigated a change in perception of the cards that would become widespread over time.

A decade later, a French occultist named Jean-Baptiste Alliette released the first Tarot deck designed specifically for divinatory purposes, the *Grand Etteilla.* This was the first deck to introduce astrological symbolism into the cards, as well as Madness, the Fool card, numbered zero.

Later decks added symbols associated with the Kabbalah and Hermeticism; in particular, the Rider–Waite deck, published in 1910 by Arthur Edward Waite, a Golden Dawn member who included many of that Order's symbols in his version. This deck, which is now commonly referred to as the Waite–Smith deck due to the fact that the artwork was drawn by Pamela Coleman-Smith, has become the basis for many others since.

Thus begins the old argument: did Waite in fact reverse the elemental associations between Fire and Air in order to maintain at least a nod to his oath of magical secrecy? Personally, I'm a stickler for Swords = Fire and Wands = Air, not only because it's the way I first learned, but also because it makes more sense.

The argument that a Sword represents Air because you wave it through the air suggests that the person making the assertion has no experience of actual swordplay, as waving a sword around leaves your defenses open and would make for a short battle. The fact that it is undeniably a martial weapon associates it in magical symbology with action, an attribute invariably associated with Fire.

Granted, a staff, sometimes used to represent the suit of Wands, can also be used in a martial context, but the leaves springing from the staves in the Waite–Smith deck represent new ideas and possibilities, attributes associated with creativity and mental astuteness—i.e., Air. The wizard's staff is generally considered a symbol of wisdom and magic rather than of war.

Having made my argument for the "correct" elemental associations, we must remember that not only are these associations effectively arbitrary additions to cards designed for games of amusement, but that symbols work deeply in the subconscious, so that someone who learns a different system will be significantly influenced by those subliminal associations.

Aleister Crowley famously released his Thoth Tarot in 1944, which he referred to as *The Book of Thoth*. He also released a book of the same title to accompany the cards. His was, perhaps, the first attempt to completely update the symbolism established by Waite. This deck, which includes symbols from the Hebrew alphabet and various esoteric as well as Alchemical symbols, has naturally remained popular with followers of Crowley's Thelema.

The artwork for the Thoth Tarot was done by Lady Frieda Harris, who author Clifford Bax introduced to Crowley in 1937, when he was seeking a suitable artist for his vision of the reimagined cards. The project was intended to take six months, but in the end spanned five years between 1938 and 1943.

The variations in artwork for more modern Tarot decks often maintain at least some of the more esoteric symbols introduced by

Waite and Crowley, though there are also many exceptions. The Morgan–Greer deck, for example, exhibits the Tetramorph (Lion, Eagle, Ox, Man) from Christian symbolism on the World card, yet otherwise is devoid of Abrahamic symbols. This is, perhaps, why I favor it for divination, as I don't have a background in those religions.

There has been much artistic license in the artwork for decks designed since the 1970s, which in some cases effectively ignore the Kabbalistic and other symbols introduced by A.E. Waite in favor of imagery from fashionable themes. Some examples include the Dragon deck, the Cat deck, Starman Tarot, and populist themes like the Disney Villains and *Buffy the Vampire Slayer* decks. There are many others, and as someone who once collected decks for their artwork, I can appreciate that the plethora of choices can be overwhelming.

Some attempts have been made to construct a Chaos deck by various small pockets of Chaos Magicians. Most of these have used photo manipulation rather than original artwork and are only known by a handful of people. I participated in one online endeavor many years ago where each card was designed by a different person, and the organizer had them printed up. We all received a copy of the deck at cost. Naturally, the result was a mishmash of imagery lacking any kind of central theme, and surprisingly lacking much in the way of occult symbols. I've never used my copy of this deck, but it was an entertaining experiment and I don't regret the small cost.

In 2014 Peter J. Carroll released his *Epoch: The Portals of Chaos*, which includes the book, *Epoch*, and the Portals of Chaos card deck, with digital artwork by Matt Kaybryn. The imagery in these cards draws from conventional symbols for the eight planets, the Ouranos and three moons symbol for Apophenia, symbols of the Elder Gods of Lovecraftian associations, and his own scrying, meditations, and invocations. It's a popular deck among Chaos Magicians, though becoming harder to find.

As a Chaos Magician, I appreciate that sublimation and the subconscious are important devices in the art of spellcraft. I'm also no stranger to the method of deliberately changing an established practice to disrupt the status quo of a situation for my own magical purposes.

For example, placing a black candle on the left and a white one on the right is well known in ritual magic based on the Golden Dawn methods, but one of my most effective spells intentionally disrupted that tradition by placing a white candle on the left (to represent purity of motive) and a blue one on the right (to represent a peaceful resolution). Sometimes a spell has to significantly disrupt the accepted state of affairs to get a desired result. In this case, the general rule was that the lawyer always wins when a person represents themselves. The success of the spell was epic, to say the least.

Most magical systems are generally very ordered in themselves, yet Chaos Magic, known for its individualistic approach to all things magical, encourages creative experimentation in ritual and recognition of the very human historical origins of all formal ritual practices. Sympathetic magic lies at the root of primitive magic and traditional forms of witchcraft, which is where a card with symbols that align to an intent comes into its own.

If a card game invented to amuse the bored upper classes can be transformed into a divination tool, it follows that these same cards can be used as a conduit of magical manifestation. After all, any spell method involves using symbols to express intent, whether it's the color of a candle, a glyph, or a constructed sigil.

With that in mind, using Tarot cards in meaningful configurations should, at least in theory, be as effective as drawing the symbols themselves, as we do in sigil magic. Naturally, this requires a close familiarity with the cards used and the meanings behind their symbols. Oracle cards might be an easier route for those who haven't delved far into the symbols on their Tarot cards, but let's stick with Tarot for the moment.

Keeping in mind that Chaos Magic encourages individualism, making personal cards is an option. It's fairly easy to find blank cards these days, or cards with printed backs and blank fronts. There are also multiple companies around that will print custom cards, so anyone who can devise their own personal meanings and depict them visually in some way has the option of creating a personal deck.

For many Chaos Magicians, this takes the form of digital art, as not everyone is good at drawing. If it's a deck just for personal use, there is less issue with professional artwork or copyright of images than would be required for a commercially available deck. A casting deck could even grow organically, using cards for sigil magic that, rather than being destroyed, as is common in sigil spells, are collected over time to be used again for similar purposes.

I've often considered designing a Chaos deck myself, using Elemental suits for the Minor Arcana like Flames and Winds to dispel any disagreements, and delving into the deeper meanings of the Major Arcana cards. Many years before I discovered Chaos Magic, I read about a deck of cards in a fiction book which included one called the Face of Chaos. The phrase was later used as the title of the fifth book in the series,[1] but it was the first mention of it that stuck in my mind. To me, Chaos represents the primordial soup of infinite possibility, which would make a card by this name equivalent to the Magician in the standard decks. Asking myself how I would depict this concept, I came up with a serpentine image, drawing from my past studies of serpent imagery in folklore and pre-Christian religions.

I've had thoughts on most of the Major Arcana cards over the years: for example, Choice to replace the Lovers, perhaps with an image of a forked path, or a sunny path working around a dark cave. Also, Disruption in place of the Tower, a concept most magicians will have encountered early in their magical paths. However, if I were designing a deck expressly for the purpose of casting spells, the imagery would likely diverge completely from the traditional associations.

Casting spells with Tarot cards can work in many forms. I have seen Major Arcana cards from the Waite–Smith deck cast into necklaces for sale on artisan sites, and I can see potential for a long-term magical energy amulet, charged in ritual, utilizing one of these, especially if it's made of gold, silver, or some other appropriate precious metal.

Working with the cards directly can be done with variations on divinatory systems—for example, choosing a card from the deck that represents the intent and working it into a ritual, leaving it on the altar for the duration of the magic. The important thing is that the symbols align with the intended result and are meaningful in context to the magician or witch.

If you have room on your altar and can safely leave cards out, arranging a card spread to encompass various aspects of a spell is a good way to establish parameters for manifestation. While there are no rules to say you cannot use the classic Celtic Cross spread for this, I don't recommend it, as there are too many separate aspects of the intent represented in a ten-card layout. Also, some of the positions signify past events, which has the potential to bring clarity to the spell, but also presents prospects where the magic could pick up misdirection if the symbols could also relate to a different situation entirely that might have happened in the past. As with any magic, it's essential to get the details right to define parameters.

If a multi-part spell is wanted, the familiar eight colors model devised by Peter Carroll is adaptable to many purposes. Casting a card layout using this system would entail placing cards at each of the eight points of the Chaostar to represent different aspects of the intent:

Red—War

Orange—Thinking

Yellow—Ego

Green—Love

Blue—Wealth

Purple—Sex

Black—Death

Octarine—Pure magic

Personally, I still feel this requires too many manifestation aspects to be able to focus a spell for specific effects, though it does eliminate the problem of past influences that might otherwise derail the intent. It's worth keeping in mind for purposes that could benefit from a broader perspective.

My own magical history goes back to a much earlier color system that has been ingrained into my psyche since my first magical experiments as a teenager:

Red—Action (to make something happen; ruled by Mars)

Pink—Love spells

Orange—To create a significant change

Yellow—Associated with intellect and intuition (good for studies or insights; ruled by the Sun)

Green—For healing and anything dealing with mundane properties (ruled by Venus)

Blue—To bring peace or fortitude in legal matters (ruled by Jupiter)

Purple—To increase power or position in a situation (ruled by Mercury)

Silver—Feminine magic (ruled by the Moon)

Gold—To address higher spiritual matters

Black—For reversing spells or dark magic (ruled by Saturn)

White—For protection or general use

As such, I am more likely to get the intended results with these associations, at least for candles involved in whatever ritual I concoct. There are many color systems these days, some involving gradient shades unknown to magicians in previous centuries. The important thing is to work with a system with personal meaning, as it is your own subconscious directing the magic.

Apart from the single-card spell, a three-card spell is a good number to bring out different aspects of the intent without getting bogged down in too many directions. One simple Tarot layout is the three-card progression of past, present, and future, reading from left to right. As Chaos Magicians, we can choose to use this in its traditional form or to adapt it to a variation that suits the spell better. Chaos Magic is all about innovation and using whatever works.

Add a fourth card, and the Alchemical Triangle configuration could be very powerful. Imagine a large triangle on the altar. At the top of the triangle, place a card representing the current situation (salt, in alchemical correlation). Moving to the lower-righthand point, speak the statement of intent and place a card representing change of direction (Mercury). Knight cards in the traditional decks are good for this, especially the Knight of Swords.

Moving to the lefthand point, place a card to indicate the nature of change desired (sulphur). Is it an important choice to be made? The Lovers. Total disruption? The Tower might be a little extreme, but the Devil also represents change to new conditions.

Then, moving back to the top point, place the fourth card over the first, imposing the outcome desired (new salt).

This sort of spell benefits from speaking aloud, to clarify the intended meaning. Any time symbols are used in a spell, a potential exists for the subconscious to interpret them differently from the conscious intent. Getting the nuances right could be crucial.

My own favored configuration is based on my progression spell method (described on my website, *http://www.jaqdhawkins.co.uk*), which uses a pentagram symbol imagined on the work space. The top point works much like salt in the Alchemical Triangle, setting the main idea of the intent, but the four following points represent aspects of the result wanted, to establish specific parameters.

The possibilities of spell construction using Tarot cards are limited only by your own imagination and ability for innovation. Inventiveness and thinking outside of the box are powerful methods for magic that attains results.

At the bottom line, a Chaos Magic approach to the Tarot is an individualized one, like everything in Chaos Magic. A magician can construct their own deck with personal symbols or use one that has personally appealing artwork. You might use it for divination or for casting, or, like me, have different decks for each of these purposes. Bringing a little Chaos into a perceptually ordered system in a magical undertaking generates extensive possibilities to attain a result of constructive change.

5

Chemognosis Redux

BY JULIAN VAYNE

Chaos Magic emerged in Europe and the Americas in the 1980s. Its formative influences included the crazy wisdom of the Discordians, the do-it-yourself ethos of punk, the iconoclastic culture of the postmodern era, contemporary science, and science fiction. From its first manifestation, in the publication in 1978 of *Liber Null* by Peter J. Carroll, the Chaos approach to occultism has often been hailed as radical. Chaos Magic itself, the name that emerged for this esoteric current, is redolent of the unconstrained, of multiplicity and of the wild. It brought with it, among other things, a professed openness to diversity and experimentation in magical practice. The revolutionary approach of Chaos Magic was articulated in *Liber Null* and, in 1982, with the publication of Carroll's *Psychonaut*, in both the form and content of the material. These books, now bound as one volume, remain essential reading for the aspiring Chaos Magician.

Chaos Magic as an emerging praxis sought a liberation from the theologically framed "magick" of Thelema, the duotheology of Wicca, and the Christian-influenced transcendentalism of Western ceremonial magic. These three systems represented the majority of the British occult scene in the postwar period, against which Chaos Magic was, in part, a reaction.

There are many examples in *Liber Null* and *Psychonaut* of the ways in which Chaos Magic sought to differentiate itself from the more religiously inclined forms of occultism. For example, in both texts the word "operator" is used to refer to the esoteric practitioner—a word perhaps suggesting a scientific, experimental, or technocratic approach to the subject of magic. This can be contrasted with other occult writing of the time, where the practitioner might be described as priest(ess), initiate, or aspirant.

This approach to magic as technology is apparent in what some have described as the "twin pillars" of Chaos Magic. The first of these is concerned with the belief system or "paradigm" the magician chooses to employ. This provides the conceptual container in which magic happens. Importantly, the paradigm can be arbitrarily selected; depending on their needs, Chaos Magicians might seek to move between a variety of belief systems or approaches. The second pillar is that the attainment of something called *gnosis* is necessary for magic to work ("gnosis" being a synonym for "trance" or "altered state of consciousness"). According to Chaos Magicians, the method of achieving gnosis should be determined by the intention of the operator and be congruent with the nature of the magical work being undertaken. We might describe these two key features of Chaos Magic as representing a concern with "setting," the literal and conceptual space in which magical activity happens, and "set," the mindset of the magician. Thus, in Chaos Magic we observe an attempt to liberate magical practice from a single setting and set, defined by the truth claims of competing metaphysical viewpoints, and open up a multiplicity of possibilities.

While the apparent iconoclasm of Chaos Magic had a postmodernist veneer, this movement in occulture can also be accurately characterized as having modernist, or even perennialist, tendencies. Supporting the stripped-back propositions of the twin pillars of Chaos Magic was the foundational assumption that, while magic may come clothed in a multiplicity of cultural costumes, it operates through identifiable core principles and processes, many of which are rooted in common human biology. Changing one's breathing, for example, whether through holotropic practice, Hindu mantras, or Anglican hymns, has comparable effects on the bodymind.[1] How the effects of changing the levels of oxygen and carbon dioxide in the blood and the altered state of awareness they create are utilized and interpreted is determined by the belief system in which the practice unfolds.[2]

What concerns us here is one specific instance of how the apparent radicalism of Chaos Magic hides a deeper continuity with the occult traditions that preceded it. This example also happens to concern one of the most potent methods for creating changes in the bodymind and thereby inducing gnosis—namely the use of psychoactive drugs.

What felt groundbreaking back in the '80s was that both *Liber Null* and *Psychonaut* assert the use of psychoactives as a legitimate part of magical practice. Psychoactive substances are repeatedly indicated as a means to attain gnosis in both texts, which, at the time of their publication, was a bold move. Indeed, the title of the second book is a word coined by highly decorated soldier and writer and, well, psychonaut Ernst Jünger in 1970, in his book *Approaches: Drugs and Ecstatic Intoxication*. While the magical practitioner is only described as an "intrepid psychonaut" once (in *Liber Null*), *Psychonaut* explores the potential use of drugs in its chapter "Chemognosis," where the reader is informed, "Chemical agents of natural and manufactured origin have always played a significant role in shamanism and magic. These substances can make various occult powers more accessible" (Carroll 1987, 147).

Despite the presence of the usual warnings and disclaimers, the inclusion of that chapter marked out Chaos Magic as different. Much other occult writing of the period rejected the use of drugs in terms redolent of remarks made a few decades earlier concerning shamanism by historian Mircea Eliade: "Narcotics are only a vulgar substitute for 'pure' trance. We have already had occasion to note this fact among several Siberian peoples; the use of intoxicants (alcohol, tobacco, etc.) is a recent innovation and points to a decadence in shamanic technique. Narcotic intoxication is called on to provide an imitation of a state that the shaman is no longer capable of attaining otherwise" (Eliade 1964, 401). (Eliade's view that drug-using shamanism is degenerate is, to put it mildly, open to question.)

The reengagement with drugs and occulture championed in *Psychonaut* correctly identifies the "significant role" of psychoactives as a continuity of magical tradition. Carroll's view is echoed in other early Chaos Magic texts. These include the subtly worded but clearly psychoactively ecstatic *Cardinal Rites of Chaos* by Ray Sherwin, writing under the pen name Paula Pagani, in 1985. Enthusiasm for the liberation of the pharmacological gnosis is more clearly articulated by a collective of Chaos Magicians writing in 1986 as the Lincoln Order of Neuromancers, or L.O.O.N., in their self-published work *Apikorsus*:

> *Of all the techniques of neuromancy, recourse to Chemognosis (drugs) is the most widespread across cultures, and in the western hemisphere particularly, one that arouses much controversy. Only those who have received medical training and can hence say from a position of authority that they do not know how the*

> *brain works, are allowed to tamper with it—through ECT, surgery and the good old "chemical cosh." While it is fine for these watchdogs to impose their will upon the brains of others, it is quite another matter for non-qualified people to try it on themselves. (Lincoln Order of Neuromancers 1986, 16)*

The desire expressed in the text, to wrestle the ownership of pharmacology away from the sole control of the medical establishment, is for good reason. L.O.O.N. further say:

> *Reaching gnosis can result, for the religiously-oriented, in "mystical experiences"—visitations by Gods, Demons, or the revealing of Divine Truths. For the magician however, the contents of such an experience are less interesting than what can be done with it—it is during moments of gnosis that sigils may be hurled; that the magician can reach through layers of space-time to manifest her will, and Gods can possess their devotees. Historically, many of the techniques of gnosis have been augmented by the use of drugs—from witches' flying ointments to the LSD & sensory deprivation experiments of John Lilly. (Lincoln Order of Neuromancers 1986, 3)*

More recent accounts of the use of chemognosis in Chaos Magic can be found in the works of several modern practitioners, including Dave Lee, and in my own writing.

These days, many writers, both scholars and practitioners, are prepared to acknowledge the use of drugs as a legitimate part of occultism. There is a plethora of books on everything from how modern witches can use cannabis in their practice to first-person accounts of intense gnostic states within indigenous belief systems. However, the views expressed by Eliade (that drug use is a recent degeneracy, and that pharmacology should be the sole preserve of medics) that L.O.O.N. railed against are still in evidence today. Take this contemporary example from the FAQ on the website of the Servants of the Light (SOL), a contemporary magical order founded by William E. Butler (1898–1978): "Drugs and magic do not mix, no matter what the experimentalists say! We regard the taking of illegal drugs or any other psychotropic or chemical substance for 'recreation' as dangerous in the extreme, and combined with occult training, this is even more applicable."

Butler wrote several popular magical textbooks, including *Magic: Its Ritual, Power, and Work* (1952), *The Magician: His Training and Work* (1959), and *Apprenticed to Magic* (1962). He studied with Dion Fortune and her Society of the Inner Light. This style of conservative, "white light" or "righthand path" occultism articulated opinions on many subjects that the nascent Chaos current reacted against. The current SOL statement accurately represents Butler's own views. Writing in *The Magician*, and struggling to castigate a range of substances he is clearly unfamiliar with, he claims, "There are many kinds of incense, and not all produce a beneficial psychic result. Hashish produces curious dream-visions, as do also *marihuana* and *peyotl* [*sic*]. All these are noxious and illegal drugs, though *Anhalonium Lewinii*, the Mexican cactus bean [*sic*] is *not* a habit-forming drug. Nevertheless, the use of any such drugs is not only an offence in law, but also an extremely unwise thing to do" (Butler 1959, 23–24).

Butler died the same year as *Liber Null* was published, which may have saved him from being outraged by the chemognostic enthusiasm of this new style of British magic.

It is interesting to note that Dion Fortune herself seemed to take a more nuanced view on the relationship between drugs and magic. In *The Training and Work of the Initiate* (originally published in 1930), she opines:

> *The difference between magic and meditation methods is the difference between drugs and diet—medicines will do swiftly what diet can only effect slowly, and in critical cases there is no time to wait for the slow processes of dietetics, so it must be either medicines or nothing. Nevertheless, drugs are no substitute for*

> *right diet and wholesome regime, and although magic enables a speedy and potent result to be attained, is only by means of right understanding and right ethics that the position which has been won can be held. (Fortune 2000, 73)*

Western occult writers like Butler who eschewed the use of drugs typically suggested that the deployment of psychoactives in magic was inherently physically and spiritually dangerous. This pharmacophobia was by no means the only phobia in evidence in the writing of this era. Gareth Knight, a prolific author on occultism, one-time member of the Servants of the Light, and magical colleague of Butler, expresses views that were disappointingly common in this period. In his book *A Practical Guide to Qabalistic Symbolism* (originally published in 1965), he states that "Homosexuality, like the use of drugs, is one of the techniques of black magic" (Knight 1978, 156).

The sinister specters of homosexuality, drugs, and black magic naturally summon to mind that giant of twentieth-century occultism Aleister Crowley, whose influence and ideas permeate the majority of Western occulture, including Chaos Magic.[3] When I was a young occultist Crowley was often regarded as a dangerous and decadent practitioner and, especially in his use of psychoactives, a rank outsider to the normative drug-free magical tradition. However, this is very much not the case in broader historical terms, for while the importance of drug use in late twentieth-century occulture was being downplayed, this pharmacophobic position was in fact the anomaly. Talk in Chaos texts of chemognosis, psychonauts, and the use of substances, from witches' flying ointments to LSD, represented a return to, and reappreciation of, the enduring role of drugs in magic.

Crowley was certainly a cutting-edge practitioner in the field of chemognosis and memorably shared his practice with an occult-curious London public in autumn 1910, where the self-styled Great Beast provided paying participants the opportunity to take part in a series of seven ceremonies in honor of the planetary forces of Hermeticism. These rites, led by Crowley, Leila Waddell, and Victor Neuburg, took place in the architecturally imposing Caxton Hall in Westminster.[4] The broadsheet advertised that the performances would illustrate the workings of an occult lodge. In fact, Crowley intended the rites not as mere spectacles but as transformational events of "self-development" for all attendees, who would witness what Crowley called "The Rites of Eleusis." To facilitate the hoped-for "ecstasy" the ceremonies included epic poetry, costumes, dramatic lighting, visual artworks, music, dance, and lots of incense. Of particular importance, alluded to by the name of the ritual sequence, was a magical brew that Crowley provided for the audience. Crowley named the ritual drink "kykeon" after the mysterious draught given, thousands of years previously, to the initiates of the Temple of Demeter at Eleusis in ancient Greece. Crowley's kykeon was a potent gnostic aid, likely to have contained, amongst other ingredients, a liquid extract of the psychedelic cactus *Lophophora williamsii* (peyote).

Amusingly, reviews of the ceremonies, published in various newspapers and fashionable journals of the day, are divisible into two groups. On the one side are those who attended but declined to drink the potion proffered by members of the Argenteum Astrum (the magical Order presiding). These reviewers generally described the event as tedious, bewildering, and strange. Some objected that they had to sit on the floor "like natives." By contrast, accounts provided by those who dared drank the magical brew describe the ritual as beautiful and spiritually powerful. Reviewers who drank the potion didn't seem to mind sitting on the floor.[5]

The fluid extract Crowley used in his brew was obtained from the pharmaceutical company Parke-Davis. Crowley was in possession of this drug from 1907 and regularly ran peyote ceremonies until 1917, attracting a wide variety of artists, writers, and bohemians. He claimed to have collected hundreds of accounts of the use of this visionary substance in his book *The Cactus*, Liber CMXXXIV. This text was lost in the 1930s, presumed confiscated and destroyed by British customs on Crowley's entry into the country after he was kicked out of his Abbey of Thelema in Sicily by Mussolini.

Peyote was the first of the "classic psychedelics,"[6] a class of substances of particular interest to the practical occultist, to enter modern Euro-American culture.[7] The first detailed modern account of its effects in English was written by the medical doctor Havelock Ellis, who also holds the distinction of being one of the first sexologists. Ellis, having been alerted to the "brilliant visions" the drug provoked by an article in the *British Medical Journal*, dropped into Potter & Clarke, the London pharmacists, and obtained specimens of the dried cactus. He published his experience, arguably the first "trip report," in 1898 as "Mescal: A New Artificial Paradise."[8] In addition to his own findings, Ellis describes the experiences of two other intrepid psychonauts. One of these was a poet "interested in mystical matters, an excellent subject for visions." However, he was, according to Ellis, impaired by a weak constitution: "He found the effects of mescal on his breathing somewhat unpleasant; he much prefers hasheesh." The poet in question was a member of the Hermetic Order of the Golden Dawn, one William Butler Yeats. Neither Yeats nor Crowley was unusual in this respect; many members of the Golden Dawn and other occult practitioners of the period used drugs, particularly hashish.

That these nineteenth- and early twentieth-century magicians were experimenting with psychoactive substances, some of which were at the cutting edge of psychopharmacology, is understandable. Chemognosis is an ancient and venerable occult technology, the deep roots of which were evinced by the name chosen by Crowley for his ceremony.

The Eleusinian Mysteries were a series of initiation rites held annually in the town of Eleusis near Athens, dating back to as early as the Mycenaean period (c. 1600–1100 BCE). These mysteries were dedicated to Demeter, the goddess of agriculture and fertility, and her daughter Persephone, whose abduction by Hades and subsequent return symbolized the cycle of death and rebirth.

At the heart of the Eleusinian Mysteries was the consumption of a sacred drink known as the kykeon. This potion appeared to play a pivotal role in inducing a state of altered consciousness, allowing participants to commune with the divine, gain insights into the mysteries of life and death, and experience a transformative gnostic state.

During the Eleusinian initiation ceremony, a reliable, repeatable mystical experience was generated for thousands of participants at a time. To explain how this was achieved, many historians have conjectured that the kykeon was a psychoactive, most likely psychedelic, beverage. Gordon Wasson (of Maria Sabina and mushroom fame), Albert Hofmann (the discoverer of LSD), and professor of classical studies Carl Ruck put forward this view in their book *The Road to Eleusis*, published in the same year as *Liber Null*. Their argued that the kykeon contained an LSD-like substance derived from the ergot fungus *Claviceps purpurea*, which grew on the grain sacred to Demeter. (Their suggestion received further support recently following archaeological excavations in 2005 at a temple shrine of Demeter in Mas Castellar, Spain, where a vase and the dental calculus of a human male were discovered that contained traces of ergot material.)

Crowley therefore stands in an ancient lineage of ceremonialists attempting to generate spiritual illumination and cultivate "self-development" using psychoactive substances.

In the nineteenth century, mind-altering drugs were also often employed in practices that one might consider closer to "operative" or "results" magic. Historian P.D. Newman describes how spirit invocation or scrying commonly involved chemognosis in that period and just who was doing it, explaining that it was:

> *[A]ccomplished with the use of prayers, invocations, and the burning of psychotropic incenses, as well as the consumption of a number of narcotics, hypnagogic, and entheogenic plants and substances. These include but are not limited to cannabis, opium, nitrous oxide, and even psychedelic fungi. . . . Some of the key players during this creative period include visionary Rosicrucian P.B. Randolph, psychic Spiritualist Emma Hardinge Britten, Helena Petrovna Blavatsky of the Theosophical Society, and especially Freemasons Frederick Hockley and his students Kenneth R.H. Mackenzie and F.G. and Herbert Irwin. (Newman 2017, 19)*

In nineteenth-century Western occult practice, chemognosis was a prevalent and quite conventional activity. Taking one name from Newman's list, we have Frederick Hockney (1809–1885), in his time a highly influential British occultist, Freemason with the rank of Grand Steward, and member of the Royal Arch. Hockney had been a pupil of Francis Barrett, author of the groundbreaking (and highly derivative) book *The Magus* (1801). *The Magus*, actually a compilation of writing primarily by Cornelius Agrippa, also recommends the use of psychoactives in ceremony. The point here is that these drug "experimentalists" are in fact noted, "mainstream" occult practitioners, far from the late twentieth-century stereotypes of the chemognostic endeavor as an eccentric or debauched behavior on the periphery of occulture.

Indeed, if we examine the wider corpus of occult literature, we find that not only are psychoactive substances used but there is evidence in some texts of an advanced understanding of pharmacology.

Writing in *Art Magic* in 1876, a welcome female voice in our story, occult practitioner Emma Hardinge Britten, describes magical rites that include "the use of Hasheesh, Napellus, Opium, the juice of the Indian Soma, or Egyptian Lotus plant." It is interesting to note that, while ubiquitous as an important symbol in ancient Egyptian art, the Egyptian Lotus (*Nymphaea nouchali var. caerulea*) was only recognized as psychoactive in the late twentieth century.

Psychoactive ointments were commonly employed during scrying for much of its history. These unguents contained ingredients including large qualities of hashish, poppy flowers, laundanum, betel nut, cinquefoil, and hellebore. Laudanum itself, a blend principally of opium and alcohol, is often credited as the invention of groundbreaking medic and magician of the German Renaissance, Paracelsus (c. 1493–1541). Paracelsus is also credited with the discovery of diethyl ether, a euphoriant and psychedelic anaesthetic, in the 1520s.

On the trail of other high-profile examples of the centrality of chemognosis we might examine the correspondence between Edward Kelly and John Dee, recorded in *John Dee's Actions with Spirits (1581–1583)*, where Dee laments that he doesn't have enough drugs for his ceremony: "I have forgotten all my druvggs (drugs) behind me. But since I know that some of you are well stored with sufficient oyntments, I do extend to viset you onely with theyr help, you see all my boxes are empty" (Whitby 1988).

While Dee and Kelly were getting high and consorting with angels, the use of psychoactives was part of the established grimoire tradition as well. *Sepher Raziel*, also known as *Liber Salomonis*, a 1564 English grimoire, gives instructions for the creation of an anointing oil necessary to see spirits in a "mirror of stele [steel]": "The third herbe is Canabus [cannabis] & it is long in shafte & clothes be made of it. The vertue of the Juse [juice] of it is to anoynt thee with it & with the [j]iuce of *arthemesy* & ordyne thee before a & clepe thou spiritts & thou shallt see them & thou shalt haue might of binding & of loosing devills & other things."

Earlier still, the astro-magical grimoire of Gayat al-Hakim, more commonly known as the *Picatrix* and composed during the mid-eleventh century, deploys suffumigations as well as ointments and confections in its materia magica. Ingredients not normally classed as drugs, such as frankincense—itself psychoactive, especially in the amounts suggested in grimoires—are of course present. Beyond such substances are a range of herbs known to have powerful gnostic potential: opium, mandrake, nutmeg, cannabis, and hellebore. These are deployed individually and in pharmacologically astute combinations to generate entourage effects, where the sum of the blend is greater than the effect of its individual parts.

Going further back in time to the Greek Magical Papyri (c. 100 BCE to c. 400 CE), beloved of contemporary ceremonial enthusiasts, we also find the use of psychoactives, which leads us right back to the period of ancient Eleusis and its psychedelic kykeon.

In short, the story of Western magic is closely bound up with the use of mind-altering substances. Chaos Magic thus represents not a radical break with tradition but an assertive return to an older lineage of chemognostic practice. Psychoactive substances are demonstrably important throughout the Western magical tradition, and if we widen our perspective to take in global occultures, we see the same pattern. While the use of psychoactive drugs goes in and out of fashion, virtually all magical traditions make use of chemognosis.

We could therefore adapt that famous diagram in *Liber Null*, which shows the lineage of the Magical Pact of the Illuminates of Thanateros, and in every box place alongside the tradition named its favored chemognostic sacrament. For Freemasonry this is alcohol (that most ancient of ecstatic allies) and, intriguingly, perhaps DMT (Newman 2017). The Knights Templar, by repute, used cannabis. Witchcraft has its hexing herbs; even dear old Austin Osman Spare, subsisting on a diet of milk stout and tea, is a moderately psychoactive sorcerer. Then we have shamanism, Tantra, Sufism, medieval goetia . . . I think you get the point.

Returning to the present, we find ourselves at the time of writing in a period in Euro-American culture being described as the "psychedelic renaissance." There are major changes afoot concerning that class of psychoactives, the psychedelics, which are typically of most interest to modern occultists. These substances were the subject of intense research and cultural exploration in the early and mid-twentieth century, until the advent of the prohibitionist period, described by the aptly insane appellative "the war on drugs." During the last decade psychedelic science and culture have been revived, and from an esoteric perspective, we are seeing remarkable things.

Current research demonstrates that psychedelic substances, correctly handled in what are often described by conventional researchers as "ceremonial settings," can have astonishing curative effects on a range of conditions, from post-traumatic stress disorder (PTSD) to chronic pain and the overcoming of addictions. Perhaps appropriately, the psychotherapeutic language used to describe the action of these substances sounds a lot like magical discourse. For instance, in discussing the use of MDMA to treat PTSD, the suggestion is that the substance (when administered in a carefully curated set and setting) activates an inherent curative capacity within the psyche. Founder of the Multidisciplinary Association for Psychedelic Studies (MAPS) Rick Doblin explains: "Our hypothesis is there is an inner healing intelligence. We all know that's true for our bodies. If you get a scratch or break bones, your body has a mechanism to heal itself. . . . There is this wisdom of the body to try to sustain itself. We think similarly there is something like that for the psyche" (Valentino 2020).

MAPS has led the way to get MDMA accepted as legitimate medicine. The success of MDMA therapy—which at the time of writing has been licensed in Australia and is soon to be licensed for use in the US and UK—seems almost magical, far outstripping the effectiveness any other therapeutic approach.

The healing potential of psychedelics is perhaps a subset of their more general capacity to enhance creativity and facilitate problem solving. There are many instances of psychedelics being used to enhance learning and generate insights, resulting in real-world applications. Numerous examples of these effects were recorded in the first wave of the psychedelic research of the mid-twentieth century and presented in texts such as *LSD: The Problem-Solving Psychedelic* by P.G. Stafford and B.H. Golightly.

While it is well known that legions of writers, artists, philosophers, and others have employed psychoactives, and especially psychedelics, as part of their creative process, there are well-documented cases of these substances helping individuals solve engineering, mathematical, and conceptual problems as well. One example of a psychedelic-inspired innovation is the polymerase chain reaction (PCR) developed by biochemist Kary Mullins. Other technologies inspired through experimentation with LSD include systems developed in the late 1960s by the members of the Augmented Human Intellect Research Center at the Stanford Research Institute: networked computers, cut/copy/paste functions, the computer mouse, bitmapped screens (the prototype graphical user interface), and hypertext. Contemporary research is also confirming that psychedelics can be used as problem-solving aids.[9]

Beyond the healing and problem-solving capacities of psychedelics, their more esoteric potentials are being investigated again today, as they were by scientists back in the 1960s.

LSD: The Problem-Solving Psychedelic also describes a range of parapsychological phenomena associated with of psychedelics in the mid-twentieth century. Contemporary scientists in the field concur that there is definitely something of interest here. As David P. Luke and Marios Kittenis state in their 2005 paper "A Preliminary Survey of Paranormal Experiences with Psychoactive Drugs":

> *The occurrence of transpersonal experiences with psychedelic substances is well attested, and several surveys have consistently found a small relationship between paranormal experiences and the use of such drugs in general.*
>
> *Even though the subjective paranormal experiences, clinical observations and anthropological reports are subject to all the usual criticisms and rebuttals that apply to non-experimental cases . . . there is a growing body of reports, rooted in thousands of years of traditional psychedelic use, that supports the notion that genuine paranormal phenomena do occur.*

Thus, we come full circle, where science (specifically, pharmacology) and magic meet. For the practicing occultist, while of course mindful of the dangers—which are often legal and not necessarily pharmacological—the use of psychoactives is demonstrably an ancient part of magical practice and has been since prehistory. Chaos Magic was and is a welcome and brave acknowledgment of this fact.

As technology develops, we are likely to see a range of new and fascinating "strange drugs" appearing in our midst that may be of interest to the chemognostic cognoscenti. Improvements in computer technology are rapidly identifying novel molecular candidates for human assay. There are new classes of substances already waiting in the wings: erotogenic drugs, cognitive-enhancing supplements, substances that provide some of the medicinal benefits of psychedelics but are themselves minimally psychoactive, orally active forms of DMT, as well as newly identified psychoactives found in both plants and fungi.

There are also likely to be an increasing number of methods for inducing gnostic states with the potency of pharmacology but without the need to ingest anything; examples include transcranial magnetic stimulation, neurofeedback systems, and virtual, augmented, and mixed reality headsets. Yet, these forms of "techngnosis" are also often themselves indebted to chemognosis. An early example of one such device is the Dreamachine, developed by Ian Sommerville and Brion Gysin following a conversation with novelist, Chaos Magician, and arch-psychonaut William Burroughs. There is also the prospect of blending technological and pharmacological methods to produce novel gnostic states (Kingsland 2019).

And so, our long, strange, magical trip comes to its conclusion. To reiterate the usual caveats, clearly the use of potent mind-altering substances isn't for everyone, let alone every occultist, be they Chaos Magicians or not. These spirits are powerful, and so must be our skills if we would consort with them. However, the potential of psychoactives when used in suitable conditions to enhance occult practice is well documented. The ability of psychoactive and especially psychedelic compounds when used sacramentally to reliably induce profound states of illumination, radical personal change, accelerated learning, and healing is clearly established.

Peter J. Carroll took a bold step in the late twentieth century by suggesting—indeed, reminding the world—in *Liber Null & Psychonaut* that drugs play a significant role in magic. While this idea may have seemed radical at the time, the weight of esoteric history was on his side. Chaos Magic cast its spell, leading to the rehabilitation of these ancient pharmacological allies in twenty-first-century occulture. Today, there are thriving communities of practice across the globe, as well as individuals and licensed explorers making magical use of psychoactives.

As a psychonaut myself, I welcome this as the return of valued and ancient spirits that were, for far too long, banished from the temple.

6

On the Causal Relationships Between Spirits and Archetypes

BY JACOB SIPES

Introduction

In this discussion I will address an ongoing debate. This debate is oriented around the question, "Do spirits exist?" There are two dominant arguments. One side says yes and supports the "spirit model" as the most plausible way to explain beings encountered during magical practice (e.g., speaking with a god during a ritual). The spirit model maintains that spirits are real. The other side says no and supports the "psychological model." This model maintains that such beings are hallucinations that express archetypes. This debate hinges on an assumed divisibility. People assume that these two models are distinct, exclusive, and, as a result, able to contrast one another.

I seek to ease the debate and mediate these models. To do this, I must contest the divisibility assumption. Therefore, my thesis is as follows:

> *Not only are both models true (at least in a limited and nuanced way), but we may also distinguish mutual influences between spirits and archetypes in practical cases.*

Once we move from the armchair to the invocatorium, these models may be neither divisible nor conflatable. We may see spirits and archetypes as distinct yet integrated objects constituting magical experiences. Thus, a third model seems required, which I will propose in this discussion.

To explore my thesis and this third model, I will make five moves. First, I will examine the spirit and archetype concepts. Second, I will examine the two models that depend on such conceptions. Third, I will examine two cases where we may see mutual influences between spirits and archetypes. In response to these mutual influences, I shall conclude the basic thesis of a third model. This model accepts that spirits and archetypes are both real and interactive. Fourth, I will discuss one key implication of the third model. Finally, I will conclude by offering a practical application of my new model. To be sure, this brief discussion is introductory and inconclusive. With that said, let us begin by examining the old models.

Models

In this section I will evaluate the veracity of the old models. I must first frame central concepts. I will begin with the word *spirit.* Spirit is an old word and has various folk meanings, but its common definition is "the non-physical part of a person which is the seat of emotions and character; the soul" (Oxford Languages n.d.). If we accept this common and conservative definition, then the old word *spirit* may be equivalent to the modern word *mind*. The mind–body problem indicates this.[1] Minds and brains are not the same. Moreover, minds, like spirits, have cognitive and emotional processes (e.g., empathetic responses). They also entail personalities and character traits (e.g., being a "virtuous" thinker). Thus, we may say that *spirit* and *mind* are interchangeable terms.

Moreover, there are two mental features that seem important but are not mentioned in the definition of spirit given above. First, language seems to be an essential mental feature. Second, the mental feature of *intention* seems to be the most important. It produces physical changes (e.g., a hunter shoots a gun to kill a deer). Minds may capture intent by language choices, no matter the language used (e.g., pictures or words). For example, a child may draw a picture of how they will eat ice cream once they return home. Thus, we may say that the spirit is not only the seat of emotion and character, but also the seat of language and intention.

Based on this conservative definition, we may evaluate whether the spirit model can be true in some meaningful sense. First let us identify the model's thesis. While there are varying kinds of spirit models (e.g., folk pagan, Catholic, etc.), the basic thesis they maintain is three-fold:

1. Some actual spirits (i.e., minds) populate the world.
2. These spirits may cause physical changes.
3. Spirits may cause changes to help us achieve desired ends.

To establish the veracity of this thesis, we must look around. If spirits are minds, and humans are mental entities with bodies, then at least *human spirits help each other achieve desired ends*. In fact, humans need such help since birth. For example, a new mother may use her intent to move her body and breastfeed her crying infant. So, we may say that the spirit model is not only true but, at least in some limited sense, also useful. It helps us predict certain human behaviors. Moreover, our spiritual nature captured by the model is a fact assumed when requesting help. We assume that, if another human can serve our interests, they must also have a mind that expresses intent and is able to influence the material world.

Perhaps we overlook this fact for a few reasons. First, people may desire to find less mundane spiritual creatures. Second, we may reject religious ideas associated with this model. Yet, the mundane *is* profound once understood. Moreover, if we reject religion, this does not mean the basic thesis of the model is rejectable by default. We may still recognize this thesis, even if humans add religious ideologies to it, and perhaps for reasons discussed later.

Next, I will evaluate the psychological model. I first address the term *archetype*. Carl Jung (1991, 4–5) defines "archetypes" as Platonic forms. Platonic forms are fundamental patterns that give rise to what we see in the world. Yet, Jung and Plato differ. Jung places forms inside the individual as potential and unconscious thought patterns. These latent patterns give rise to myths, fairytales, and sacred symbols. Jung proposes that these archetypal expressions organize the mind as it develops. He refers to this developmental process as *individuation*.

However, Jung also admits that archetypes are *hypothetical*. We can only know them by analyzing *personalized* myths, fairytales, and sacred symbols. To discuss archetypes, we must reference private ideas. Thus, we must also use phenomenology and hermeneutics to interpret something individual.

Last, if we are to define archetypes, we need another feature. Archetypes are *aspirational*. Mythic stories and powerful symbols often inspire strong urges. They seem goal-oriented. For example, we may read J.R.R. Tolkien's *The Lord of the Rings* (an explicit myth and fairy story). This English myth inspires deep urges for beauty and enchantment. It moves many different people to try similar things, like new foods and world travel. Perhaps more importantly, *self-myths* entail strong urges regarding personal development. For example, it is common to hear a Protestant express how they strive to "conform to Christ." They wish to become more like the Christ they imagine (e.g., Christ the loving healer, Christ the heroic martyr, etc.). They conform their behaviors and life choices to this dream. They ask themselves, "What would Jesus do?" Many have developed positive characteristics through this aspirational process (e.g., heightened empathy).

Based on the above definition, we may evaluate whether the psychological model can be true in some meaningful sense. First let us identify the model's thesis. While there are varying kinds of psychological models (e.g., postmodern pagan, nontheistic, etc.), the most basic thesis they maintain is threefold:

1. The mind (i.e., the spirit) is inhabited by unconscious patterns of thought.
2. These patterns express themselves as personal myths and sacred symbols.
3. These myths and symbols may cause changes in mental states and behavioral choices.

To establish the veracity of this thesis, we must look at our aspirational behaviors. If spirits are minds, and humans are mental entities with bodies, then we may say that *archetypes help human spirits achieve developmental progress*. In fact, we see this begin during early childhood. For example, a young boy may encounter a story about a brave knight who saves a fairy princess. Afterward, the child imitates the knight in dress and behavior. As they grow, they may even come to follow an aspirational question, "What would the brave knight do?" Boys like this tend to become men who pursue "knighthood" as a life path. Geoffroi de Charny was such a boy.

He grew into an outstanding medieval knight. After years serving his king and winning wars, he also wrote an influential book about Western chivalry in which he outlines how knights should achieve developmental progress and strive toward ideal knighthood. His lived dream was captured by the pen for all to read. In sum, people like Charny conform to a personal myth which captures developmental urges and drives growth in a specific direction. Thus, we can claim that actual fairytale knights and wizards exist. People with big dreams, and the guts to chase them, make myth a reality.

Therefore, we may say that the psychological model is not only true, at least in some limited sense, but essential for growth. In fact, *self-mythology* seems to be a fact assumed when stating who you desire to become. Like the small boy who says he will become a great knight, we assume that, if we can and do become a specific and desired person in the future, this imagined future must entail some story of that future self which compels us. Charny dreamed of becoming an *ideal* knight that behaved a specific way and valued specific things. A distinct self-image pulls us forward. It guides our choices. It disrupts our current habits and beliefs. It also motivates new ones that may be drastically different from the old.

Perhaps people overlook this fact due to the desire to believe in a permanent self. Perhaps we wish to disassociate with psychological forces that challenge ego authority. We may even hide from the power of our aspirations. After all, they imply deep change, risk, and surrender to a "greater" self. Most people hate feeling small, unstable, and inadequate. Regardless of these feelings, we may still recognize the basic thesis which incurs them. Myths and sacred symbols, *and the aspirations they entail*, may still be experienced—even if, like an army, we evade surrender by engaging in guerrilla warfare against them.

Given the above, we may conclude that both models are true. Yet, mages may still try to conflate spirit and archetype. A psychological mage may attempt to interpret all spiritual creatures (e.g., a summoned "demon") as mythic hallucinations, akin to a schizophrenic break. The spiritual mage may attempt to interpret all magical visions (e.g., a dreamed angel) as spirit communication, akin to a divine revelation. However, as I will now address, such contrasts and conflations seem problematic. I now examine four possible interactions between spirits and archetypes that demonstrate the issue.

Relationships

Causal relationships help us make distinctions. We can identify how objects are not the same due to how they influence one another. For example, a person may break a glass bottle and cut their left hand. They may use the left hand to glove the right, so that the right hand can handle the glass safely and treat the wounds found on the left. Though they belong to the same body, and are even the same category of objects, they influence each other in different ways, sometimes to different effects, and at different times. If these body parts are distinguishable by such interactions, even if they belong to the same body *and* body part category, the same may apply to mental objects that have far less in common.

I will now outline several causal relationships between spirit and archetype found in practical life. To do so, I will explore two cases that represent common experiences that seasoned mages may encounter. The first case demonstrates how an archetype may influence development changes in a spirit. It concerns the calling of a young wizard.

Jamie's Case: Relationship One

Presume Jamie, a young academic, enters a local bookstore. He discovers a book titled *Liber Novus.* The cover art shows a red dragon eating itself. It also shows a wizard. The wizard image provokes Jamie, so he thumbs through the book's contents and becomes engrossed. Afterward, he feels compelled to look for other books like it. Jamie finds another book titled *Liber Null & Psychonaut*. He discovers that this book is a magical grimoire and feels that he must take it home.

While walking home, Jamie thinks about the wizard image. It fascinates him. At home, he researches historical wizards. This search involves delving into wizard etymology, history, world practices, and fantasy representations. He digests countless wizard stories. Yet, he feels dissatisfied by just reading them. He wishes to practice wizardry, and test if this satisfies the new urges. So, he pulls the grimoire from his bookshelf. After several exercises, Jamie discovers that, while practice is vital, this too is not enough. Instead, a new and wondrous *ideal* begins to emerge in his mind. An ideal wizard begins to appear to him in nighttime dreams, waking imaginations, and artworks. It calls him forward. By interacting with and contemplating wizard myths, he uncovers a powerful urge. He must *become a philosopher-mage* (the ideal wizard of his dreams). He must understand and master the very foundations of magic. He then commits to this life path and makes necessary changes to achieve this ideal (e.g., he changes his academic disciplines, daily practices, etc.).

If Jamie is a spirit that has a body, and his *spirit is populated by archetypes* that express themselves as personal myths and symbols, we might say that an archetypal self-myth has spontaneously emerged. It has influenced distinct apsirational changes. There are even drastic changes concerning life choices. He develops new habits, skills, and values which conform to this self-myth. Thus, we see a causal relationship between archetype and a spirit's individual development.

If this is the case, we may also see a similar archetypal influence on a *social* level. To explore this influence, let us continue Jamie's case.

Jamie's Case: Relationship Two

After spending time growing on his own, Jamie feels he must begin a new search. To follow the wizard image deeper, he feels he must seek someone else that shares his aspirations. Perhaps a mentor will know what inspires Jamie and help him achieve greater growth. So, Jamie thumbs to the back of his now-worn copy of *Liber Null & Psychonaut* and, after finding the author's contact details, sends an email. To his surprise, the author has a magic school. Jamie joins. The young wizard and his aged mentor spend time working through the coursework.

After graduating, Jamie appreciates that he and his mentor differ in certain approaches and views. Yet, he also appreciates that, despite these differences, both share a general wizard myth. They both seek to become an ideal *scholar-mage* who attains comprehensive knowledge and technical mastery. Though their myth details differ, general "wizardry" is a shared myth, one that brought the two together due to shared aspirations and mutual development. If Jamie and his mentor are minds (i.e., spirits), and aspirational archetypes populate both minds, then a shared myth *motivates their interactive development.* Shared archetypes may bring spirits together for mutual gain.

Justin's Case: Relationship Three

While we have explored the influence of archetypes on spirits, we may see the reverse. Spirits may influence archetypal expressions like personal myths. Myth changes may have significant ramifications. Presume that Justin, a middle-aged mage, has embraced "Celtic polytheism." He adopted Senuna as a tutelary deity. He even refers to himself as a druid and belongs to a local Grove. Justin venerates his goddess and serves his Grove as a daily discipline. In short, this mage holds a powerful personal myth.

Now let us presume Justin disagrees with recent decisions made by Grove leadership. This leads to his dismissal from group activities. He performs rituals and seeks Senuna's guidance. But she remains silent. With nowhere left to turn, he explores different groups and their shared myths. One group is a local Catholic club that reads mystical literature and practices contemplative methods. The only deity they venerate is Jesus the Logos. Their methods intend to produce personal gnosis and guidance.

After spending time with the group, he reads the new literature and performs the prayers. He has new magical experiences. As a result, he adopts a new personal myth. Before, Justin pursued being Senuna's perfect disciple. It was his pleasure to honor his goddess and protect the natural world. Now he views himself as Jesus' disciple. Justin feels he must conform to the new god's example. So, he volunteers with charities, begins attending weekly mass, and forgives others. His new habits involve imaginative meditations. Justin imagines Jesus and himself merging into the same person. As a result of this discipleship, family members report that Justin's character has drastically changed, and in certain ways, it has greatly improved.

As represented by this case, we may abandon one myth, its corresponding archetypes, and the behaviors it inspires. We may engage with different myths and their corresponding archetypes, and explore a different impact on our spirit. By doing so, we may change what archetypes express themselves. By doing so, we alter what inspires certain behaviors and, as a result, how we impact the world. In short, while archetypes may influence spiritual change, we may also see the reverse. Spirits may influence archetype changes by altering the personal myths they adopt. Moreover, we may see a similar causal relationship on a *social* level. To explore this influence, let us continue with Justin's case.

Justin's Case: Relationship Four

Justin shares his spiritual visions with the group, confiding that Jesus the Logos has provided personal teachings. The main teaching is this: to achieve cosmic harmony and benevolence, the group must make concerted sacrifices in the form of group fasts and community service. The group disperses with the promise that each member will consult the Logos about this private instruction.

The group reconvenes. Each group member shares their own visions, which confirm the need for fasts and community service. The group agrees that the Logos inspired these sacrifices. They decide on a first fast which lasts thirty days. In addition, they organize and execute a park cleanup project. Afterward, the group share their subsequent visions and their interpretations of how their sacrifices pleased Jesus. Group members were shown symbolic images like planets aligning according to a new sun. They were also shown future results of these sacrifices (e.g., future children receiving the benefit).

After completing more fasts and service projects, each member shares how these experiences changed their views. As a group, they agree that the Logos of the World is far more active and concerned about small affairs than previously believed. Moreover, they accept that contemplative prayers are not enough for discipleship and conformity. They cannot be like the Logos if they only sit idle. Instead, group actions that conform to the Logos are also required. In sum, Justin's private revelation produced great social change. Inspired by this private revelation, subsequent revelations and experiences reshaped the spiritual group.

Therefore, we can say that if the Catholic group are spirits, and archetypes populate this group, then Justin's influence produced an *interactive development of group archetypes* (or at least the development of their expressions). Changes at the personal level (e.g., having a new vision of Jesus) inspired new and concerted group changes. Spirits, as a collective, may alter what myths they obey due to individual influences. Thus, we see a causal interaction between a spirit group and its archetypes as mediated by group members. One spirit may help another "attune" to different archetypes, or at least different archetypal expressions and influences.

Conclusion

If the cases of Jamie and Justin represent certain common experiences, we may now draw a key conclusion. We may conclude that causal relationships between spirits and archetypes act *as a dynamic dialogue.* Spirits and archetypes exchange specific influences. Thus, I offer this conclusion as the central thesis of a new model, which I shall call the *dialogue model.* This model may be codified by three premises:

1. Both spirits and archetypes exist, at least in some limited sense.
2. Both spirits and archetypes interact in causal ways.
3. Interactions may cause a range of changes in practical life.

As we can see, this new model encompasses the basic theses central to the old models. It also sets the groundwork for future premises and research. In addition, it captures causal influences that may benefit the mage.

On the one hand, archetypes may change spirits through personal myths and sacred symbols. Moreover, archetypes may change groups of spirits through shared myths. Shared myths entail shared aspirations. Such shared aspirations may lead to enhanced results through interactive development. In sum, archetypes seem to exert an important influence on spiritual development, and on the world due to inspired behaviors.

On the other hand, spirits may change archetypes by engaging different mythic representations (e.g., recounted experiences, books, songs, etc.). Spirit groups may also change shared archetypes. Such changes in myths may even alter group development. This might help explain why collective behaviors change as group myths evolve. In sum, spirits seem to exert a causal influence on archetypal expressions, and subsequently, through these changes, alter what inspires group behaviors.

As a final remark about the dialogue model, we cannot ascertain its veracity if we let ideological assumptions blind us to the subtleties of our experiences, or at least to the potential experiences that human spirits can encounter. If experience and careful analysis guide our magic models, *and not theistic, nontheistic, or atheistic worldviews*, spirit and archetype conflations may be unjustifiable. More importantly, we cannot attempt to *use* these causal relationships to produce benevolent benefits if we make such assumptions. A mage may disagree with my proposal, and yet, proceeding as if the argument is true, they may still apply these ideas as a test. Provided they are at least receptive to experiences that contradict their current views, such a use may lead to surprising experiences that liberate their spirits from ideological mires. With this model proposal now closed, I will explore a key implication of the dialogue model.

Implication

To remind the reader, the question at the heart of the ongoing debate is, "Do spirits exist?" I have answered in the affirmative by offering conservative theses about the spirit and psychological models. As shown by the preceding conclusion drawn from the cases considered here, we may say at least that some spirits exist, and that archetypes play an important role in their growth. However, if the above is true enough, then this suggests another possibility. We might see causal relationships, at least of the types outlined in this discussion, demonstrated between *nonhuman spirits and their myths (and sacred symbols)*. Let us exercise our imaginations and explore a hypothetical example.

Presume that there exist some nonterrestrial life forms on a distant planet. Also presume that they have minds (i.e., spirits) that display the same characteristics as those of humans (e.g., intent and complex language use). However, they possess silicon-based bodies. (Humans have carbon-based bodies.) These aliens are seven feet tall, translucent, and emit a static charge. Now let us say they have learned to control the displays of these electrical charges and use them as a language medium. (In contrast, humans have learned to control carbon dioxide streams to form a language medium.) Using these discharges, the aliens tell stories of their gods and dream of their future selves. If we can say that the Platonic forms active within one human may be the same as or akin to those active in other humans that exhibit shared mental characteristics, the next logical step may be to acknowledge the possibility that nonhuman life forms that exhibit familiar mental characteristics might also be influenced by the same or similar archetypes.

Moreover, if we can say that archetypes may exert a social force, where they bring together human spirits for mutual benefit, then perhaps archetypal expressions (e.g., sacred symbols) that attract other human spirits *may also attract nonhuman spirits*. Those spirits that share comparative myths and symbols might assist in our spiritual development (as we assist in theirs). The dialogue model might be generalizable, and not bound only to human spirits. Thus, I close this section by proposing a practical rule that anticipates the possibility of nonterrestrial spirits which share our "Platonic forms." This rule is:

> No matter the body possessed, *spirits* of a mental feather will flock together.

Application

To use the rule, we must raise a more general question. How can a mage use the dialogue model? Above I have outlined a model that proposes a causal interaction between spirits and archetypes. This interaction suggests a first use: submission. If archetypes can influence individual *and group* behavior, then those who make use of such an influence have a powerful tool. They can conform to select archetypes, receive inspiration, develop new behaviors, and impact the world in good ways. A mage should exercise self-awareness and deep prudence when obeying such myths. Seek seasoned council.

The model also suggests a second use: direction. If spirits can change archetypes, then this suggests that the mage may shape how archetypes influence people. They can do so by becoming *a master of archetypal expressions*. For example, the mage may become a storyteller, or become a painter offering sacred symbols. By mastering some fine art as a vehicle for magical praxis, the mage may guide the archetypal forces acting on a social group. As an inspired artist, they may initiate, inspire, and inform others through publication (as I am doing now). This influence is exerted even if the recipient is not aware of the guiding hand. Thus, the mage, as a human spirit, may engage the archetypal forces in themselves and others, using myths and symbols, and so *guide* the secret dialogue.

By applying the dialogue model, we thus discover a third use: integration. By guiding the interaction, a mage can change the power dynamic that helps to govern their lives. They can put archetypal influences on their terms and not be subject to unregulated compulsions. Archetypes may compel behaviors we do not want. By shaping archetypal forces (while still tactfully allowing those forces to shape them), mages may balance different internal directives.

On the one hand, archetypes mediate the needs of other archetypes, like the "higher self." These forms are eternal and inexhaustible. They compel behaviors that may even contradict bodily needs. For example, we could work all night to obey an ideal when the body requires sleep. On the other hand, the mage mediates the needs of embodiment. Human spirits have bodies with biological urges, such as reproductive and survival instincts. These instincts compel behaviors that may even contradict our self-myths. For example, a person's sexual orientation may contradict a purity ideal. The mage, as master of archetype expressions, may integrate the needs of the spirit and those of the body and, by doing so, shape these interests to achieve a full life.

As a final note about mastering archetypal influence, one of the best ways to guide the secret dialogue is to combine all three uses. To do so, the mage must *become a conduit* of archetypal influence. Like the boy who becomes a knight, the mage may turn themselves into a living myth. To become a living myth, they must achieve great things and do legendary deeds in pursuit of a personal ideal. Others will then look to them for inspiration.

How do we achieve these great things and do these legendary deeds? By conforming to a specific personal myth, often referred to as the "true self" or "higher self." Your own personal higher self will always be specific and unique, even though it begins with or takes ingredients from something more generic. Self-myths have parents, yet the children are individuals.

For example, Christians may claim they are "Christ-like" and share the life of Jesus as a group myth. Yet, each Christian conforms to the Christ figure they imagine. One imagines Christ the poor healer. Another may imagine Christ as a holy warrior. Mother Teresa was just being "Christ-like" when she was inspired to feed millions. The Crusader was just being "Christ-like" when he was inspired to slaughter the Jews of Jerusalem. The generic Jesus becomes part of a personalized story, and through the Christian's conformity to *their* Jesus gives rise to a unique Christ in the world. The finite children of an eternal god are always born distinct.

As a second example, our young wizard found his face in the "wise man" myths of different cultures. He surrendered to the myth of his higher self by becoming the philosopher-mage he dreamed of. He conformed to it by taking the necessary steps to develop and strive for this ideal. Pursuing his ideal required constant sacrifice, patient evolution, and hard work. While ideals are nonachievable (perfection always requires more work), the higher self still compels constant, and sometimes drastic, positive changes in the direction of itself. Thus, the wizard committed to getting as close as possible to his ideal for the rest of his life. Achievements and changes produced by such ever-growing conformity eventually inspire others.

In sum, generic myths and sacred symbols act as a culturally conditioned mirror through which the individual finds their face and follows it. By consciously conforming to their dreams of a "true" or "higher" self, mages may achieve positive transformation *and* social impact. The dialogue model supports that a spirit expressing such archetypal force will provoke the archetypes shaping other spirits. Thus, the mage comes to control who they attract as the rule applies:

> No matter the body possessed, *spirits* of a mental feather will flock together.

I now end this discussion by offering some advice. To change the spirits you attract, *human or otherwise*, you must first change who you want to become. As a spirit, you must engage your personal myths. Direct your archetypal forces and let them direct you. Dream your true and higher self. Give birth to the god you are. As Rumi says, "You are what you seek."[2]

Thank you for reading, and be blessed.

All the best,
The Faraway Philosopher

7

A Path into Animist Sorcery

BY AIDAN WACHTER

An Animist Dream

In the deep dark of the deep-time past, all the Powers you knew. You knew them on a cellular level, as you knew yourself on a cellular level. Words were vaguely on the horizon somewhere as you turned ideas over in your dreams and painted them on your skin. Words would come, but not quite yet. Names had not been invented, and signatures had the nature of scent and textures in your mouth, under the calloused pads of your hands.

Later, names came, and along with them walked language. The arrival of a spoken reality. Dreams had voices. Voices carried dreams between dreamers. Dreams became co-creative with the Others,[1] *those with whom we shared inner dream space, and also with other dreamers.*

What arose from this moment was the ability to speak shape to the dreaming. We call this sorcery, and we call this magic. Conceptualizing not only what is but what could be. To speak-sing it into being.

This speaking, this telling, brought about a different relationship with the previously unnamed and always ineffable Powers: the spirits, Others, the Dead, and the Gods.

Thousands of generations passed in this new shape of the Dreaming. We (because the past didn't go anywhere; we are our ancestors as they are us carried forward) spent long ages hunting, gathering, and communing with these Powers.

Our bodies know this. Understand it implicitly. It's wired in our neurology and the cells of our flesh. We are creatures of blood, bone, and spirit. Sorcery is the language we speak to the Others.

This is the Dream.

This is the Story.

How I Got Here

I found the Chaos current a few years after I began reading about magic, when a friend gave me a copy of *Liber Null & Psychonaut* in 1987. Carroll's writings resonated with me. They seemed more pragmatic than mystical, looking at different "kinds" of magic more as approaches that could be used rather than anything like a doctrine of spiritual truth.

It seemed to me that magic, if it worked, should be more like auto repair than some faith-based practice. In time, I would find myself in a very different place, but that was my sense of things at twenty years old.

I was a diligent student. I meditated, did a ton of sigils, and practiced various banishing rituals, invocations, and other solo and group rituals. In the coming years, I lived in three houses that were semi-public working temples. I trafficked with Chaoists, Thelemites, ceremonial magicians, Christian magicians, feminist witches, LGBTQ witches, hoodoo and conjure practitioners, and those involved in various African traditional religions.

I also had some experiences with uninvited possessions and a few moments where I walked into extremely liminal spaces while wide awake, usually out in nature. What I learned shaped my actions and led me to mentally sort the magical practitioners I met into two primary groups.

The first group practiced magic because they *believed* the world could be a very different place from the one they were experiencing. They came to magic to find out if this was indeed possible. They *hoped* to find a world of magic and actively sought mystical or spiritual experiences.

The second group came to practice because they had *already* experienced mystical or spiritual experiences. On a profound level, they *knew* there was more to the world than the purely material reality perceived by the people around them. They practiced to gain a better handle on experiences that were already present and active in their lives.

I was firmly in the second group. The explanation for why the world was as it was that was offered to me in the late twentieth century was remarkably flat and lifeless. It ignored and denigrated as primitive bodies of knowledge that had survived for ages. It categorized vast ways of being and modes of perception that had carried our ancestors' genes to us across time as the superstitions of fools and idiots. It seemed to do so, to my mind, mainly because they were inconvenient to the control strategies used by the powers of church, state, and industry.

I practiced, and I learned.

Experimenting with versions of the *Mass of Chaos B*, an invocation of Baphomet, taught me to enter controlled possession states intentionally. I modified that framework to work with other entities and archetypes; in time, those presences changed me.

This was all in a roughly five-year "program" that might now be classified as the "fuck around and find out" school of magic. It worked! In time, it worked consistently. However, *how* it worked was often a bit unpleasant. Around this time, I found *Visual Magick: A Manual of Freestyle Shamanism* by Jan Fries, which struck a strong chord in me. I largely abandoned all the "use anything that works" and Discordian sorts of approaches that seemed to be rising in the Chaos Magic world.

These included, more specifically, what one might call the "toys and comics" schools of thought among many Chaoists I knew.[2] I understood and accepted that cartoon characters could be workable archetypes of power, but the combo of silliness and what could be called "ironic detachment" became off-putting. It ran counter to something I'd been learning about magic: while it isn't required, at some point, sorcery begins to work better if approached with sincerity and honesty. Treating magic as you would have magic treat you is one way to think about it.

In time, I came to some conclusions about both my own practices and magic more generally:

> *Magic is a way of perceiving, being in, and interacting with the world. Most (if not all) humans are born with it. Like other human traits, it manifests along a spectrum from seemingly no ability to mind-bogglingly high ability.*
>
> *Magic is a skill. Everyone who can get out of their own way and practice a lot gets better. This is true of playing guitar, cooking, or sorcery. You might not excel at any of them, but you will improve.*
>
> *I found that I operated most effectively within a spirit model.[3] I perceive spirits; they communicate with me, they guide me, and at times they fuck with me. We operate in a generally symbiotic relationship.*
>
> *It seemed that many people who have difficulty performing effective magic (in other words, their workings don't tend to work) are often committed to a model, an approach, or a set of beliefs that don't suit them. This is a bit like wanting to fix your car while being determined to use woodworking tools. This methodology is unlikely to succeed.*
>
> *Most people understand the spirit model of magic innately. It is rooted in ideas of animism, which is the belief that much of*

the environment is alive, or animate. This belief or experience takes many forms. Some believe all that exists is alive and in some way sentient. Others believe that many things have what is known as an indwelling spirit. From this perspective, not every stone is alive and aware, but some are via this indweller. I believe this is an accurate model and one wired deeply in human neurology. However, you don't have to believe it for good results. It's "true enough to be useful" even if it isn't objectively true. In other words, we know how to make the system work, even if we are incorrect about its underlying mechanisms.

A Spirit Model for the Spirit-Agnostic

The spirit model can be abstracted into interesting patterns and alternative views that make it more digestible for those who find the idea of spirit work uncomfortable. Spirits can be viewed as psychological phenomena, projections that we can work with most effectively if we behave "as if" they are external entities. I know a lot of practitioners who work with spirits in basically the same way I do who hold this view. Their take is that it provides a useful framework for engagement, even though it is not an accurate rendering of how the world actually works.

We can see versions of this approach in diverse forms, from Tsultrim Allione's *Feeding Your Demons* to Ramsey Dukes's *Little Book of Demons* and even the school of modern psychology called Internal Family Systems. All of these have in common an understanding that our minds can best communicate with some "parts of self" if we can *talk* to them. Thinking about a thing is different from conversing with that thing.

When we expand on this beyond the internal mindscape, we can begin shifting from a simple spirit model into a form of animism. If we are willing to consider that there may be "parts of self" that we can interact with "as if" they are independent beings, how much of a stretch is it to consider that there may also be "parts of the collective unconscious" that really do operate as independent beings? And how much of a stretch is it to imagine that there may be "parts of the collective unconscious" that persist from those who are now excarnate? Is there space in this hybrid spirit/psychological model for spirits the size of gods? Or for the presence of the ancestors, as allies or as ghosts? For indwelling spirits of land, place, or object? Is there space for the Others, of whatever form or nature?

Another Story: A Biological View of Animism

Magic, sorcery, and witchcraft are all key parts of what makes you human. They lie below the level of purpose, meaning, in this case, your sense of the intention behind your life. They operate at the level of function, which here refers to what you, or beings like you, do within the system. Purpose is intentional and personal, and function is structural, shared between similar parts of the larger system. Purpose is tied to a sense of identity, where function exists on a cellular level.

Magic, in all its forms, is a way of inter-being with the Field and its inhabitants.[4] *The work of magic is the work of opening (which often seems more of a remembering of) channels of power and awareness of the interface between oneself, the Field, and the Other(s). This can happen when one seeks mundane ends just as well as lofty spiritual attainments.*

The practice of magic allows for a breakdown of systems of internal oppression caused by generations of control doctrine that underlie systems of human civilization. However, it can also be used to reinforce these systems, to calcify those structures into stone and bone, seemingly permanent, seemingly impenetrable.

As a magician, witch, or sorceress, no matter how much you focus your work on improving your state personally, you are also helping to strengthen the connections and inter-beingness of the system. Consider these connections as neural networks. They are like the synapses of your brain firing and wiring to create new memories, music, and stories. They are like motor control neurons that fire to contract a muscle fiber to raise your arm, move your eye, or turn a key in a lock. They are vast networks of structure, energy, agency, and intention, and they link humans into the wider net of the living Field. You aid the Field by intentionally interfacing with it. Magical practice builds stronger connections, allowing greater and clearer passage of information. This information has a guiding effect on both the system and your experience within it.

Looking at what worked for me magically, I saw some patterns. My practices worked best if:

I truly believed the outcome I desired was possible.

I left lots of room for how the results manifested, meaning I didn't try to micro-manage the path of manifestation.

I needed (rather than wanted) the outcome but wasn't needy in how I asked for it.

I made a lot of offerings.

Using the largest, widest lens possible, I applied Occam's razor—which suggests looking for simple solutions rather than getting all hot and bothered by adding complexity—to the "stories" around all I knew about magic. The result was that vast generations of my ancestors were some version of animists. They believed implicitly in the existence of the Others, be they human, the Dead, land spirits, animal and plant spirits, spirits indwelling places, stones, streams, homes, trees, and tools. They lived with and were surrounded by these beings, which they viewed in some way as people. I thought, "Why not take this as thousands of years' worth of hard-won wisdom and run with it?"

In time, I came to believe that the idea that all these beings were radically different and distant from us came much later, after the arrival of beings of the priest class, the self-proclaimed intermediaries between people and the Field.

Here is yet another story that explains my biological view of animism.

You are a spirit being that inhabits a body. You share that body with many other beings. These beings can be bacteria or viruses. Mitochondria.

You inhabit a larger body—the Earth, your home body—and you can be viewed as a small being within that body, like bacteria, viruses, and mitochondria are within your own. Or perhaps, if it's more comfortable for you, you could consider yourself a cell. You have a role to play as a cell in that larger being. You may help make up a filtering organ, like a liver or kidney, or a structural part, like a femur. A small part of a retina, or a nerve, a myelin sheath.

Is this a small thing? Does it feel uncomfortable to be a small thing? Is it hard to de-center yourself to the degree where you are able to say, "I am a small part of a larger part of a still larger whole whose nature is vast and incomprehensible from where I am, with the tools I have"? I assure you, it is not a small thing! You are a crucial aspect, be that of sight, of removing toxins from a body, of helping a creature to walk upright on two or four legs!

Dream Logic

This world, this life, is a dream. You are a dream within a dream, and you yourselves are dreaming—yourselves, as you are many. Many cells make you up, and many spirits and ancestors have supplied you with your component parts and component natures. Stardust and sea kelp and spores and molds and fishes you are. Elk and clear streams and dead moons.

This is a dream. All of reality is the shape of the dream called the Field. Many of what we have been taught as "laws" and "rules" can be thought of as the current dream logic of the world. This dream logic can and does shift over time. What was once known as truth is now laughed at. This does not mean it wasn't once true. The dream has shifted. And yet.

And yet the deeper reality, the deeper dream, still runs deep in the background, as deep as the rivers of the Underworld. Can you feel them? Do you hear their waters as they rush among the stones of the World Below? Can you feel them as they course through your veins and bathe the inner workings of your structure?

The deep dream is the realm where the deep mind roams among what many consider to be the *archetypes* that I know as the *Others*. Allies, friends, Powers attentive, hostile, or seemingly oblivious to my being.

Earlier, I spoke of those who *sought egress into* a magical reality and those who already *lived in one*. Here are two more possible points of view.

The first is those who center themselves as *independent agentic Powers*. The second is those who view themselves + the Others + the Field as *interdependent agentic Powers*. These two views often clash, as it seems like a crucial aspect of identity for most people: Do you believe in only yourself, or do you believe in yourself *and* (insert your own favorite term here)? Meaning, yourself in the total context of *being in community with*?

Another way to consider these perspectives is this: Do you view magic as the interaction of your own consciousness and will with a largely mechanistic world, where you are more or less pushing buttons and pulling the levers through your workings? Or do you view magic as an interaction between your consciousness and other consciousnesses, perhaps an ultimately conscious universe or Field?

This second view of interrelated, interlaced, enmeshed consciousnesses is the root of animism, as I understand it. We operate *with* the Field, not simply *within* it. We operate as at least semi-autonomous spirit beings *with* other at least semi-autonomous spirit beings. And if we believe that the entire Field is active and responsive, we are having back-and-forth conversations with it and the Others. We're not so much instructing or ordering the universe to meet our needs and desires as nurturing a kind of symbiotic relationship that, by its nature, breeds communally beneficial outcome states. A healthy ecosystem that nourishes all of its components.

While my lifetime of experience has me "knowing" in every way that matters to me that I live in a relationship with independent beings and a conscious living Field, this is not required to practice magic in this way. It seems enough to practice "as if" this is so.

Let me explain that a bit. Imagine that it is your strongly held position that you are the center of the universe. You are the central point, and all spins around you; all is moving relative to you. One way to practice is to project this idea into the world forcefully. In truth, for many people, this works and works well. You ask, and you receive. There's no actual sense of reciprocity or symbiosis. We could call this the Sovereignty of the Will model, perhaps.

That's how you view the world. But you look at historical spirit-based animistic perspectives and decide to experiment with them, to try them on for size. You start to make offerings to your allies, ancestors, and the various spirits you know have been believed in and perceived in other cultures or at other times. You still have a spirit-agnostic view, but you don't *act* like it. You act as if you live in a spirit-filled world, surrounded by allied spirits and supported by the vast collection of ancestors whose lives led to yours.

You begin to ask for guidance from these Powers on the best ways to craft your rituals, shape your spells, and guide your mundane actions to support them. When things go well, you make offerings. When things go poorly, you make offerings. Because the offerings are not a transaction—you aren't paying for your results with coffee and incense. The coffee and incense are part of your relationship, nourishing the surrounding allies and the Field itself.

Magic is, in part, a collection of arts for focusing *intention* and *attention.* Attention is *where we look.* Intention is *what we need or desire.* I believe that this animistic approach works so well because it plays into the strength of our neurology as it has evolved and developed over time. In other words, for many practitioners, as soon as they begin to act *as if this is true and natural,* parts of the deep mind begin to open up, to flower and bear fruit. When we say, "Okay, now light a candle for your spirit allies," even if your rational mind chafes and wants to rebel and call bullshit because you are "above all that superstitious nonsense," your mind and body *know* this way. It's in you. You are here because of a deep ocean of humanity that *lived* this experience. You are at least in part *made* of this.

A Path into Animist Sorcery

The application of all these things I've talked about is incredibly easy. While the overall systems are complex, the on-the-ground work is extremely simple. Everyone understands, on some level, how to have a good relationship with another being.

Step one: Invite the powers you want to work with into your space. This is a process of attraction rather than compulsion. For all that is good and right in the world, don't try to *use your will* to *make* this happen! Be friendly. Expect good things. Be relaxed and open. Let what happens happen.[5]

Step two: Make offerings for these Powers that you have invited in. This is a process of *acknowledging* them. This is where you say, "Hey, I know you are here, aware and active in this space. Thanks for coming by!" Water, candles, sugar or honey, bread, incense, and song (as in, you sing, hum, or play music using your voice or body—it's OK if it's not objectively good!) are all excellent offerings for the allies. Think of these as fuel for the spirits, allies, and Powers. This is perhaps easiest to understand in the case of candles: as the candle burns, it releases energy stored in its wax and wick as light and heat. The allies can feed off this energy. With this approach, they receive nourishment from all we give to them. Perhaps add some material bases—spirit houses—to help them establish themselves in your zone and anchor them in your life and the spaces where you primarily make offerings.

As you make these offerings, clearly define who they are for. The language I use is something like this:

I make these offerings
I give these gifts
To all those spirits
Who aid and guard me
To all the ancestors
Who aid and guard me
To all the beautiful powers
Who ward and watch over me
Eat, be nourished, and grow stronger!
Let there be peace between us all of our days.

Now, you may have to adjust this if, for whatever reason, you are getting unfriendly visitors at your table. This is uncommon, but not wildly so. It is more likely if your practice is heavily based on bossing such beings around, as is traditional in many systems. If this is the case, begin with making offerings to one "strong protector" spirit that either you already work with or who calls to you. Ask this power to protect your space and workings.

In my experience and that of most I know, this process of offering starts something akin to mutual fermentation with the particular Others who walk with us. They tend to self-regulate, pushing out any harmful "bacteria" or "viral" elements (the more troublesome powers and spirits) as they grow stronger and our relationships with them deepen.

Results tend to show up if you repeat this offering for sixty days or so, with a frequency of every other day or better. In this early stage, consider adjusting any other magical work you do and, perhaps even more importantly, your internal, written, or spoken dialog to match or, at the very least, not run counter to the work you are doing with your offerings. In other words, if you do your offerings and then talk shit about the process, how it's bullshit, how it isn't working, how you don't believe . . . you aren't doing the work. You are actively working against your success. Correct your actions in all areas to match your intentions at the altar. This is where the idea of belief as a tool comes in: for the duration of your experiment, you will behave in all ways as if all of this is the clear-as-a-bell known truth about how the world works. For this, you can use your will!

At some point toward the end of this sixty-day period, when it feels natural and unforced, you can ask for help. Yep, help. Assistance. You want these powers, whose sphere of influence differs from yours, to help you. Ask nicely. Not on your knees, not begging, just nicely. This works like this: you do your regular offering process, and at the end, you ask in a normal voice for what you want help with. "Hey, I know y'all know I'm trying to get a job. It would be excellent if you would help out. Here's what would help most, but you do what makes sense to you. Thank you." And that's that.

Now is a good time to start writing about the kind of work you desire and perhaps doing some other magic for it. But do all of this as if you *know* that it's the allies that are going to hook you up. You are doing this work to help them out and lighten their load. You are obviously (because you are a smart cookie) going to do absolutely everything you can do on the mundane level to make this new job happen. Getting help with a resume, asking employed people you know about known openings, not drinking excessively all the time, improving your personal hygiene . . . doing all the things that you know will logically make you more employable.

Whenever something beneficial happens, it would be *wise* to thank your allies. You don't have to give them total credit, but give them some. You meet some cool people? Someone helps you out? Bring it to the allies: "Hey, I like that woman I talked to at the cafe. She seems really smart and kind. If you had anything to do with that, it was awesome. Thanks." "Jimmy asked his boss about openings, and even though they didn't have anything right now, it was super cool. If you were involved with that, I really appreciate it."

I want to make something clear. When you approach life and magic this way, it's possible that you are training aspects of your mind, and those mind parts are all "Hell yes, finally some appreciation!" New synapses are formed, and they start getting on board with manifesting a better reality. That's not what I think is going on (or at least not entirely), but it doesn't matter what you or I think. What matters is if it works.

You can expand this approach as far as you like. Might your home have an indwelling spirit? Where might it live? The kitchen? The front porch? What happens if you start feeding it? How about that park you always feel good visiting, or the seashore, or your front door? This kind of magic is about building direct relationships with the world around you, seen and unseen. It's not very flashy and doesn't require much in the way of study or monetary investment. It is also probably not suited to everyone. But for those who run the experiment and find that it does work well, it has the potential to transform your practices and life. That is, after all, what magic is really for.

I offer these words to you, their reader. May your magic be nourished and grow stronger.

8

Thread Theory: A New Chaos Approach

BY IVY CORVUS

A Brief Look at Chaos Magic

One of the most fundamental pieces of a Chaos Magic practice is the idea that one's belief in something gives it power—and, in turn, can shape one's reality. Chaos Magicians often use the phrase "belief as a tool for manifestation," which indicates that one's belief is a potent tool for magical work. If one believes in a God and uses its power to manifest their desires into the material realm, then it is their belief in the God that gives the magical working its power. It does not matter whether that God objectively exists. What matters is the Chaos Magician's belief in the God, and that subsequent results manifest in their material world.

When discussing how to *do* Chaos Magic with my students, I find it difficult to give instructions. This is because Chaos Magic is a philosophy and a way of perceiving how magic works. It is not a

magical system with set rules to follow. When someone asks, "How do I do Chaos Magic?" this is the equivalent of someone asking, "How do I do pantheism?" Or rather, "How do I do animism?" These aren't philosophies that one *does*, these are philosophies that one *believes*. A magician or an occultist practicing Chaos Magic believes that their belief gives power to the magical working. In this moment, I could walk out to my backyard, scoop up a handful of dirt from one of my garden beds, and spit in it, and this would now be magical dirt infused with my magical DNA.

And so it is; so mote it be.

One of the more challenging aspects of Chaos Magic is strengthening one's ability to believe in what one is doing. If one can hold a handful of dirt in their hands and proclaim that this is now magical dirt infused with one's energy and intention, how does one get themselves to believe this to be true? Chaos Magic may seem simple in theory, but it can be much more difficult in practice.

A common technique of Chaos Magic is a process called "paradigm jumping," also known as "belief shifting." This is a process of intentionally adopting and discarding beliefs depending on the practitioner's magical objectives. By fully immersing oneself in a chosen belief system, the Chaos Magician can tap into the power of their subconscious mind during ritual to create change and manipulate their reality. I would take this a step further and express that frequent and successful paradigm jumping also helps to decondition the ego, remove brainwashing and societal limitations, and strengthen one's ability to believe in anything one sets one's mind to—including imbuing a handful of dirt with magical properties. But how does one begin to paradigm jump? Is there a more efficient way of strengthening one's belief in their ability to manipulate experiential reality?

"Nothing is true, everything is permitted" is another common phrase in the Chaos Magic community. Many will be familiar with it from the *Assassin's Creed* video game, and it is often attributed to German philosopher Friedrich Nietzche. However, the origin of this phrase has various possible sources, including an attribution to Hassan-i Sabbah, a military leader from the eleventh century. This expression is often used to convey a perspective on subjective versus objective reality, existential freedom, and Chaos. Is it possible that all religions are both true and untrue? Is it possible that monotheism, polytheism, pantheism, and atheism can exist under one construct of reality? What if the Chaos Magician could more effectively paradigm jump between these beliefs?

All of these concepts will be explored as we discuss what I have identified as *thread theory*. This includes the idea that all perceptions of reality are both true and simultaneously untrue, as dimensions of thought overlapping one another. We will explore the concept of threads connecting individuals to each other, to the collective, and to different realities of existence. Thread theory also provides the practitioner with the tools to manipulate our reality by cutting and manipulating certain threads tethered to us.

Many of my students of Chaos Magic have expressed that they find it difficult to shift their perspective; to jump paradigms and believe in something new; to deconstruct their minds, remove societal brainwashing, and cultivate the most efficient magical practice possible. Thread theory provides a paradigm that helps practitioners reduce their rigidity when it comes to the dogma they subscribe to, while also validating differing personal realities among various Chaos practitioners. But before we discuss the components of thread theory, some foundational knowledge must be established first.

Models of Magic

In the dead of night, a witch lights a candle and casts a spell. A psychic medium works with spirits to channel divine messages.

A seer experiences premonitions of the future. Do these practitioners commune with divinity where spirits manifest objectively, or are they unlocking facets of their subconscious minds, realms yet uncharged by the scientific community? What constitutes the driving force behind this magical power? Numerous models of magic exist, each delineating different perspectives on the mechanisms underlying magical and paranormal phenomena.

The spirit model, for example, suggests that the power of magic comes from spiritual entities. Practitioners may invoke or work with spirits, deities, or other supernatural forces to achieve their desires. In this model, spiritual entities are objectively real regardless of the practitioner's belief in their existence. The energy model suggests that the power of magic comes from energy. This model conceptualizes that there are subtle, nonphysical energies that exist and can be manipulated to bring about change in the physical or spiritual realms. The psychological model suggests that the power of magic comes from the practitioner's mind. The combination of ritual, symbolism, and belief affects the practitioner's psyche and perception, causing physical changes in the external environment. Rather than believing in spirits, as those who subscribe to the spirit model do, the practitioner working from the psychological model believes in projections of the subconscious mind. The meta model of magic suggests that the power of magic is unknowable, as we can only subjectively experience reality. Therefore, those that subscribe to the meta model may use any other model of magic at any time, as they see fit.

There are many other perspectives and models of magic that exist to explain how magic and paranormal phenomena occur. Chaos Magic is usually associated with either the psychological model or the meta model of magic. Thread theory arguably adopts the approach of the meta model of magic, but provides more of a unique system to operate within and a set of techniques the practitioner can follow if they choose.

Subjective Versus Objective Reality

"Nothing is true, everything is permitted."

To say that "nothing is true" is to acknowledge that one is unable to experience the world objectively. Because one experiences one's existence as a human from a subjective perspective, one cannot say that spirits, Gods, inherent moral facts, or various metaphysical concepts are objectively real.

Subjective reality and objective reality represent two different ways of perceiving the world. Objective reality refers to the external, independent existence of things, regardless of an individual's perception or interpretation. It exists independently of personal opinions, feelings, or perspectives. Subjective reality is based on personal opinions, feelings, and interpretations. The idea that a human can't have a completely objective experience stems from the inherent subjectivity of human perception and cognition.

There are many reasons why achieving objectivity in human experience is impossible. To begin, human perception is inherently subjective. Our senses, such as sight, hearing, touch, taste, and smell, are filtered through individual experiences, biases, and interpretations. What one person perceives may be different from what another perceives. Humans also have cognitive biases that affect the way information is processed and interpreted. These biases are often unconscious and can influence decisions, judgments, and perceptions, leading to subjective experiences. In addition, cultural and social factors play a significant role in shaping individual perspectives. Cultural norms, values, and societal expectations influence how people perceive and interpret events, making objectivity difficult. This includes societal brainwashing, where individuals within a society are influenced or manipulated to adopt certain beliefs, attitudes, values, or behaviors that align with the prevailing norms and ideologies of that society. A person's emotional involvement also introduces subjectivity into how they perceive the world.

All of this is not to say that humans cannot strive for a more objective and impartial perspective. Self-awareness, critical thinking, and a willingness to consider alternative perspectives can help alleviate some of the subjectivity humans experience. However, complete objectivity, free from all personal influences, is an elusive ideal. Humans cannot detach completely from their vantage point enough to be able to claim that any sort of objective reality is true. I'm not suggesting that an objective reality cannot exist; rather, I'm suggesting that we can't claim any one particular objective reality to be true.

To say that "everything is permitted" is to acknowledge that because nothing is true, and there is no objective reality that can be observed from the human experience, the individual has full creative freedom to create the reality they desire. If one were to open their mind to the possibility of all things being true and simultaneously untrue, like layers of different timelines, dimensions, perspectives, and beliefs overlapping one another, one could achieve anything. Immediately, I sense that there may be some readers skeptical of my last statement, wondering how far one can go in manipulating reality. No, I am not insinuating that one can climb up to the top of their roof, jump off, and fly because they expect to be able to and believe that they can. We will address this argument later, after establishing a few other players in the thread theory game.

The Many-Worlds Interpretation Versus Thread Theory

The *many-worlds interpretation* (MWI) is a hypothesis in quantum mechanics that suggests that every possible outcome of a quantum event occurs in a separate and non-communicating parallel universe. This is just one of several quantum mechanic interpretations, and it is a topic of heated debate between physicists, philosophers, and even Chaos Magicians. MWI was first proposed by physicist Hugh Everett III in 1957 and remains theoretical to this day. However, it does raise interesting questions about the nature of reality.

One of the key components of MWI is the concept of branching universes. According to this interpretation, when a quantum event occurs with multiple possible outcomes, the universe splits into different branches. Each branch realizes one of the possible outcomes, creating one of numerous parallel worlds. Thread theory is similar in that it recognizes multiple different realities of existence. However, MWI states that these parallel universes are entirely separate and do not interact or communicate with each other. Each universe, according to MWI, evolves independently, maintaining its own set of physical laws and conditions. This is not the case for thread theory.

Imagine this scenario: a couple have built a home and a life together after being married for many years. The first partner is an atheist and does not believe in spirits. This partner has never experienced paranormal activity, nor do they believe in any sort of psychic phenomena. The second partner believes in spirits, has experienced paranormal activity at some point in their life, and does believe in psychic phenomena. Both individuals exist together in a joined, mundane realm. However, these individuals also have threads connecting them to different dimensions of experiential reality. In the first partner's dimension, spirits do not exist and they do not affect the individual's human experience. In the second partner's dimension, spirits do exist and they do affect the individual's human experience. Both perspectives are true and valid in their dimension, while also being false in the other.

A strong argument has already been established for the subjectivity of what humans experience. Thread theory suggests not that we exist in parallel universes that are entirely separate from one another, but that we can exist in multiple dimensions at once. That is, these two individuals can exist together in a mundane realm while also existing in their differing spiritual realities. The first partner exists in a realm where spirits do not exist; the second partner exists in a realm where they do.

Thread theory also suggests that when one recognizes which threads they are tethered to, they can cut old threads and tether new threads to change the reality they experience. An atheist who has a desire to believe in spirits and experience paranormal activity could go through the process of cutting old beliefs and tethering new ones. An individual raised in a religious cult who wants to shed limiting beliefs could do the same. My students of Chaos Magic have asked me countless times how to overcome the religious indoctrination they were subjected to that no longer aligns with how they wish to see the world. Residual fear can be cut from the individual, with new beliefs tethered in its place. An individual could even shift themselves into a reality where a handful of dirt is magical. Recognizing the threads one is tied to opens up innumerable possibilities.

So What Is Thread Theory?

Now that a framework has been established to discuss thread theory, what is it exactly? Thread theory explains that there are energetic threads connecting every living entity to a collective pool of consciousness, to each other, and to different dimensions of reality. I use the term "dimensions" here to refer to the different facets of what we perceive to be reality. Every living entity has many threads connected to them, some stronger and more stubborn than others, and we can manipulate them by cutting unwanted threads and tethering new ones. Techniques for how to do this will be discussed later in this essay.

The idea of threads, strings, or cords made of energy is not a new concept. Some may be familiar with the idea of energetic "strings" from string theory. I am not a physicist, nor will I ever claim to be, but my understanding of string theory is that it suggests that the building blocks of the universe are not point-like particles but tiny, vibrating strings. As these strings vibrate and change shape by twisting and folding, they produce effects in tiny dimensions. Such effects may underlie the behavior of the whole of reality.

Some may also be familiar with the idea of "cords" connecting living beings to one another as a sort of energetic bond. It has become increasingly popular in the witchcraft community to perform cord-cutting rituals when one wants to sever the relationship or emotional connection with another. However, this idea is not solely found in witchcraft. Many ancient healing traditions, such as traditional Chinese medicine, Ayurveda, and various indigenous practices, emphasize the flow of energy within the body. The concept of energetic cords may have evolved from these traditions and others.

For example, an individual may have an energetic bond connecting them to their partner, their family members, their friends, and even their mailman. Every exchange and interaction between living beings creates a cord that connects these individuals together. The stronger the relationship, the thicker the cord, for better or worse. This could potentially explain how some can sense when a close friend is about to call, how a mother always seems to know when their child is in danger, or how two people can't ever seem to move on from each other regardless of how much time has passed or where they find themselves. Though strings and cords could be appropriate terminology, I employ the term "threads" to express the idea that we are engaging with a more expansive and intricate design, like the threads woven into a tapestry.

Another player in the thread theory game is the concept of a collective unconscious, introduced by Swiss psychiatrist Carl Jung as part of his analytical psychology. This refers to the idea that there is a shared, universal pool of memories, symbols, and experiences that all humans are connected to because of our common ancestral past. This collective unconscious is separate from one's personal unconscious, which consists of an individual's unique experiences and memories.

Jung proposes that individuals can access the collective unconscious through dreams, myths, and creative expressions. Themes or symbols that emerge through these mediums often reflect the archetypal elements residing in the collective unconscious. For example, people from different cultures may experience similar archetypal symbols—say, the image of a snake, often associated with transformation or danger—in their dreams, reflecting a shared symbolic meaning.

What is compelling about the idea of a collective unconscious is that various anthropological studies have identified historical instances where human societies, separated by long distances with no contact with one another, independently developed very similar practices and beliefs. Examples include similarities in mythology from culture to culture, and similarities in death rites, initiation rites, solar worship practices that personify the Sun as a deity, and more. In an era when indigenous communities could not have contact with distant communities on separate continents, we observe that they still developed similar practices and beliefs. How can this be?

Up to this point, it has been established that there is potential for an individual to have threads connecting them to other living beings, to the collective unconscious, and to different belief systems that have a direct impact on their external reality. We have also established that these many dimensions, or facets of reality, aren't separate worlds or timelines. They are overlapping, as one can exist in many at one time.

Going back to our original example of a household with two people existing together yet experiencing very different realities: an atheist and a monotheist or polytheist can both be correct and simultaneously incorrect. Each individual has a thread tying them to their partner, their home, and their mundane life, but each individual also has a thread tying them to their belief in gods and spirits (or lack thereof), which ultimately creates the reality they experience. So, how does one go about shifting one's perspective? If an atheist had a strong desire to believe in and experience supernatural phenomena, or if a religious cult survivor wanted to remove dogmatic or limiting beliefs acquired through religious indoctrination, how would they go about this? We will explore various techniques in the next section.

Paradigm Jumping Techniques

For some, simply knowing that a perspective like thread theory exists is enough to help shift their perspective to the reality they wish to experience. When I first started experimenting with the idea of paradigm jumping, it was incredibly difficult for me to shift my mindset. I wanted to believe in (and experience) spirits, but I was a long-time atheist and thought this would never be possible for me. I always found myself a bit jealous of those who could easily experience paranormal phenomena. After shifting my perspective from the idea of ultimate truths to what I have identified as thread theory, I was able to remove the blockages and limitations that had held me captive for so long.

In my experience, the first jump is often the hardest and requires the most dedication. Now, after many years of successful paradigm jumping, I no longer need to follow strict, regimented techniques to get myself to believe in something new. However, for a beginner, a strict and regimented approach may be necessary, especially when working with stubborn, long-held beliefs. Below are some exercises tailored for individuals who are starting their journey with paradigm jumping.

I recommend mixing and matching these techniques to customize a practice that works best for you. It may be beneficial to perform a combination of these techniques simultaneously. I can say with confidence that they do work. I now experience inexplicable paranormal phenomena regularly, despite experiencing none when I was an atheist and did not believe in their existence. I was able to shift my reality to experience a dimension where spirits exist, and so can you.

Thread-Cutting Visualization

This exercise involves visualizing a thread tethered to your body, energetically cutting it, and tethering a new thread in its place. Begin by finding a comfortable and quiet space where you won't be disturbed. Sit or lie down in a relaxed position, allowing your body to sink into your chair or bed. Close your eyes and take a few deep breaths, inhaling deeply through your nose and exhaling slowly through your mouth. With each breath, feel yourself becoming more and more relaxed. I would recommend looking into various methods of breathwork to find a technique that works best for you. An altered state of consciousness must be reached before advancing to the next step. I recommend achieving an alpha brainwave state for this exercise. I don't have the space here to fully explore brainwave states, but I encourage you to read Mat Auryn's book *Psychic Witch* to learn how to enter altered states of consciousness prior to ritual.

Once you have achieved an altered state of consciousness, bring your awareness to your body. Visualize a thread extending from your body, connecting you to the belief you want to remove. This thread may appear as a thick rope or as something more delicate, whatever resonates with you. Take a moment to observe the thread and notice how it feels. What does your reality look like with this belief held in place? Trace the thread back to the source. Where does this thread come from?

As you continue to breathe deeply, imagine yourself reaching out with your energetic scissors. With intention, visualize yourself cutting through this thread, severing the ties that bind you to this belief system. Feel the release as the thread dissolves away, freeing you from any negative attachments or energies.

Now, envision a new, vibrant thread of light appearing before you. This thread represents what you truly desire to be tethered to. With a sense of determination and purpose, reach out and grasp this thread, feeling its energy pulsating through your being. Visualize what your reality looks like with this new belief held in place. What can you do with this new belief? What does it feel like? What do you experience with your five physical senses?

As you tether this new thread to your body, feel yourself becoming filled with its empowering energy. Allow it to wrap around you like a protective cocoon, nurturing and supporting you fully. Visualize yourself bathed in its light, radiating strength and autonomy. Take a few moments to bask in the sensation of this newfound connection, knowing that you are now aligned with what truly resonates with you. With each breath, feel yourself becoming more grounded and centered in this new energy.

When you are ready, gently open your eyes and return to the present moment, carrying with you the clarity and empowerment of this meditation. Know that you have the power to release old, limiting beliefs and to tether yourself to the energies that inspire you.

Feel free to record this meditation by reading it out loud in your own voice, and to play it back to yourself. I recommend performing this meditation daily for a duration of three months. Some readers may be shocked that I suggest such a long period of time. To put this in perspective, if you are attempting to cut the tether to a limiting belief that you have held on to for the past ten years (or more), it is unlikely that you will be able to sever this thread completely after performing one meditation on a random Saturday night. This process takes time and dedication.

The first time I attempted to cut the thread to my limiting beliefs surrounding religious indoctrination, I performed this meditation every day and documented my progress to see just how long it would take. I anticipated it would take many months, but after approximately sixty days, I experienced a significant shift. Some may experience faster results than others. Some may need to perform this meditation daily for much longer than three months, depending on how stubborn and long-standing the belief is. However, I tend to recommend three months because it is usually around the two-to-three-month mark that I have experienced these shifts.

Subliminal Messaging

Subliminal messaging refers to a technique of conveying information—a thought, an idea, or a stimulus—to the subconscious mind, without the conscious awareness of the individual receiving this message. Implementing subliminal messaging by recording yourself speaking aloud an intention phrase and playing the recording while you sleep is a very simple way to implant new beliefs into your subconscious mind.

For example, someone wanting to experience more paranormal activity might use the intention phrase "I see spirits." I recommend writing intention phrases in the present tense, as if it is something you are already experiencing. So, instead of saying, "I wish to see spirits," you would say, "I see spirits." You could also add additional intention phrases, such as "I am open to seeing spirits" or "My ability to see spirits is growing stronger every day."

After choosing your intention phrases, record yourself speaking these phrases out loud. You can record yourself repeating the phrases or simply put them on a loop to create a longer track to listen to. Subliminal messaging tracks can range anywhere from a few minutes to a few hours in length. I recommend a track that is at least thirty minutes long; the longer the track is, the better chance you have for success. You can also consider adding some ambient music to the background of the track you have created.

For subliminal messaging to work, the intention phrases must be implanted into the subconscious mind. This can either be done through deep meditation (by reaching a theta brainwave state) or by listening to the track while you sleep. Listen to the track you have created every single night for one month. Assess your progress, and adjust accordingly. I recommend attempting this technique for three months before expecting long-lasting results.

Audio Sigils

The next technique involves creating what I identify as an audio sigil, or an audio signature, to chant during a meditation. Similar to the idea of mantra meditation, you are chanting certain sounds in an altered state of consciousness that invokes a certain state of being. An audio sigil differs from a traditional sigil in that it is a sound signature created from a phrase rather than a visual representation. The reason why I do not recommend traditional sigils here is because I find that visual sigils are best for quick magic—not necessarily a technique for creating long-lasting new threads. The effect of sound and vibration on the mind is outside the scope of this essay, but I encourage readers to investigate it nonetheless.

To create an audio sigil, or audio signature, first define your intention phrase. For this demonstration, I will use the same example, "I see spirits." To begin, remove the vowels so that you are left with only the sounding consonants:

I SEE SPIRITS
S S P R T S

Vowels are removed to turn the intention phrase into a series of sounds, rather than full words, which is more effective for attaining altered states of consciousness. As you can see, "S S P R T S" still keeps the integrity of the words intact. Even without the vowels, the sounding consonants still make the sound and shape of the words in your mouth. Allow yourself to slip into an altered state of consciousness as you chant the sounding consonants, repeating them without pause:

S S P R T S S S P R T S S S P R T S S S P R T S S S P R T S S S P R T S

Continue chanting as you move deeper to a theta brainwave state. Slur the sounding consonants together until they are rhythmic sounds escaping your mouth. When performing this technique, the sounds may eventually slur to the point of nonsense. Continue chanting until a theta brainwave state has been reached. I recommend at least a thirty-minute meditation every day for one month. Then, assess your progress and adjust accordingly. If you do not see changes after three months, try a different technique or evaluate your opposing subconscious beliefs further. Opposing subconscious beliefs will be discussed later in this essay.

Why Paradigm Jump?

Why might one want to shift dimensions and experience a different reality? Who cares? Why would someone go through all this work, practicing something every single day for three months or more? Yes, that does seem like an intensive time commitment. As discussed previously, it takes time to decondition our minds and remove limiting beliefs that have been held in place for potentially several years. There are many reasons why I believe paradigm jumping is necessary to become a more effective magician.

If one can shed limiting beliefs, they have the potential to become more enlightened and more effective in their magical practice. As an analogy, consider a relationship that you once thought was beneficial for you but that, after it ended, you came to understand did not best serve your interests. Have you ever been in a relationship with someone you thought you were going to marry someday? Maybe the relationship ended, and you were left heartbroken, crying, and reminiscing over the good times. We tend to romanticize events and relationships until we can gain enough separation to experience some clarity. After the pain fades, years later, you look back at that relationship and realize it wasn't nearly as good as you thought it was. In retrospect, you might even feel grateful you didn't marry that person. Perhaps you were not really compatible, or they were holding you back in your personal or professional life.

Shedding limiting beliefs is necessary even when you do not see them as limiting in the moment. It allows you to experience personal growth by opening you up to new opportunities and possibilities. Shedding limiting beliefs also improves the magician's confidence and self-esteem, allowing them to pursue their goals with stronger conviction. When we let go of limiting beliefs, we become more resilient in the face of challenges. Instead of being held back by negative thoughts, we are better able to adapt and overcome obstacles. Limiting beliefs can also act as barriers to manifesting our desires. By releasing these beliefs, we can attract more of what we truly desire.

When I began frequent and regular paradigm jumping, something happened to my brain. To this day, I still cannot quite articulate what it feels like to someone who has not experienced this. Frequent paradigm jumping broke down barriers in my mind that I never knew existed. The more I jumped, the easier it became to believe in anything I set my mind to. I experience little to no resistance when I walk out to my garden bed, scoop up a handful of dirt, spit in it, and proclaim that this is now magical dirt with magical powers. It *is* magical, because my will is strong enough to create a reality where this is true.

Subconscious Beliefs

It's important to note that when we are assessing and manipulating our own threads, we are not only working with the conscious mind. We are also working with the subconscious mind, which holds beliefs we may not be aware of. These beliefs can include self-sabotaging behaviors and thought patterns, such as:

* I do not have the power to change my reality.
* I fear change, and I fear this new belief system.
* This is all imaginary make-believe, and I am acting childish.

It is necessary to deconstruct these subconscious beliefs to ensure that your new threads can attach properly. Journaling about your limiting beliefs and thought patterns is an excellent way to discover subconscious beliefs that may hinder the success of your practice. Start a regular journaling practice that incorporates the following prompts:

* Where did these limiting beliefs come from?
* What past experiences may have influenced these beliefs?
* What messages have you received from authority figures or society about these beliefs?
* What reoccurring patterns or themes do you notice in your life about these beliefs?
* How do you talk to yourself internally?

* What would it mean and what would it look like if you let go of these limiting beliefs?
* What fears or insecurities surface when you consider making these changes in your life?

The involvement of the subconscious mind is why it is crucial to enter an altered state of consciousness prior to performing the paradigm jumping techniques discussed previously, such as thread-cutting visualizations, meditating with audio sigils, and subliminal messaging. When entering an altered state of consciousness, the conscious mind relinquishes control and the individual is better able to implant these new beliefs into the subconscious mind.

So, Can I Just Fly? What About Physics?

If one can manipulate their reality to be anything they want it to be, can we perform actions that seem impossible, such as defying the laws of physics? Not necessarily. I would like to remind the reader that thread theory suggests you have many different threads connected to you at any given point. This includes threads that not only connect you to your belief systems, but also to other individuals and the collective unconscious.

The collective unconscious itself carries memories and rules that are common to us as a species. You may find yourself in a game of tug-o'-war with what you desire to believe against the limitations of the collective. Imagine sharing a brain with an entire human race that believes we can't live in a dimension where flying is possible. (Let's pretend flying is a metaphor for all things that seem nonsensical or impossible.) Trying to defy the laws of physics by attempting to fly off your rooftop would be an attempt to overthrow the hive mind and defy the rules that keep us tethered to this version of reality.

Threads that tether us to other people also pose limitations. Have you ever struggled to believe in something that your partner directly opposes? People in our lives who are tethered to us can influence our ability to tether ourselves to a new idea or concept.

The rules of our biological bodies in this dimension of reality can also be limiting and provide challenges for us to overcome. To clarify, I am not suggesting that escapist thinking is a solution to real problems and challenges in life. Such thinking is highly problematic, and not a healthy alternative to participating in this reality we have created as a collective. In no way am I insinuating that one can simply will away a physical disability, for example, with the power of shifting realities.

That said, theoretically, if we were to change the perspectives of the entire human consciousness, could flying then be possible? Who knows? I am certainly not qualified to answer that question. However, thread theory suggests that it is not about changing the rules of this dimension, but about shifting to a new dimension entirely, with different rules in place.

Interesting research by neuroscientist Dr. Anil Seth suggests that our perception of reality is a controlled hallucination. I hope readers will catch my sarcasm with this next statement: maybe one day, centuries in the future, we will be able to access sufficient brain power to make flying possible. I don't foresee this happening anytime soon.

Do Spirits Care What You Believe?

An argument opposing thread theory is the idea that spirits don't care what you believe. They exist regardless of your belief in them. It doesn't matter if you declare your belief in spirits; they could care less about you and will continue existing either way. I have two counterarguments that address this idea.

First, to claim that spirits exist regardless of one's belief in them is the same argument as someone asserting that their God is the one true God. It is the same as a Christian claiming that there is only one God, the same as a polytheist claiming that there are multiple Gods, a pantheist claiming that there is "God-like" energy interspersed within everything, and an atheist claiming that there is no God at all. As discussed in the section on subjective versus objective reality, we cannot claim that there is any ultimate truth. Because humans are only capable of experiencing reality subjectively, there is no credibility in claiming that something ultimately does or does not exist.

My second counterargument comes from personal experience. When I was an atheist, I wanted so deeply to believe in spirits. I wanted to experience them, talk to them, and live amongst them. However, due to my conditioning, I was unable to believe in their existence. For years, I tried various rituals to invoke and evoke spiritual beings. I sought out teachers, spent nights in haunted locations, and tried various summoning techniques with no success. During that time, I never experienced any sort of paranormal phenomena, which was disappointing to say the least. I had experienced paranormal phenomena when I was a child, but as an adult, I was unable to re-create those experiences. It wasn't until I started utilizing the techniques outlined in this essay that I was able to shift my perspective and begin believing in spirits.

When I finally achieved this shift in perspective, I started experiencing paranormal phenomena regularly. These were experiences I could not rationally explain with my critical mind. I began experiencing spirits moving objects in my physical environment, speaking to me, and showing themselves for brief moments. For a while, I thought I might be going insane. Hearing voices and seeing things that weren't there before was cause for concern. However, upon experiencing physical objects being moved, I realized this wasn't solely in my head. A curtain was moving in the shape of a being with the window closed and no draft in the air. A single book was pulled out of the bookshelf while the other books stayed in place. There simply was no rational explanation for these events.

Now, if I perform the exact same rituals I tried and failed as an atheist, I can achieve success. My methods did not change, but my beliefs did. If spirits are real regardless of what we believe, then what explains my experience? Why would spirits begin to interact with me after I began to believe in their existence? Was my deep desire, as an atheist wanting to believe, not enough? I suspect that when I paradigm jumped into a belief in the existence of spirits, I tethered myself to a reality where spirits *do* exist.

As an experiment, I decided to test this theory further. I went through the process of tethering myself back to the belief that spirits do not exist. After successfully reattaching this thread, I no longer experienced paranormal phenomena. This confirmed my belief that the threads we are connected to do shift the dimension of reality we experience. There can be a world where spirits do and do not exist, depending on which dimension the individual resides in.

What if I Believe in Paranormal Phenomena but Still Don't Experience Them?

This is another question I get quite frequently from students studying the occult. My response is usually this: I believe polar bears exist, but I have never seen one in person. I also don't make any effort to travel to the right places to see a polar bear in the first place. I cannot expect to see a polar bear without putting in the effort needed to make it so.

If you believe in paranormal phenomena but still haven't experienced them, that may be more indicative of the efficacity of the techniques you are using to experience these phenomena than it is indicative of whether they exist or not.

Isn't Thread Theory Just the Baader–Meinhof Phenomenon?

While there are similarities between some aspects of thread theory and the Baader–Meinhof phenomenon, there are also significant differences. The Baader–Meinhof phenomenon, also known as frequency illusion, is a cognitive bias where once something is noticed for the first time, it seems to show up everywhere shortly afterward. This phenomenon occurs due to selective attention or perception. The brain becomes more attuned to the thing that has been noticed, which leads to a perception that it is occurring more frequently than before. For example, if someone wants to buy a blue car, they might start noticing blue cars more frequently. This isn't necessarily because blue cars are showing up more often; they are simply noticing the blue cars more often than they did before.

Thread theory is a more complex and multifaceted concept. It illustrates that individuals are connected to various dimensions of reality through energetic threads, which influence their perceptions, experiences, and interactions with the world. These energetic threads do not simply direct our attention in one direction or another, affecting what we notice. Rather, they represent deeper connections to belief systems, collective consciousness, and even different dimensions of experiential reality. Thread theory also suggests that individuals have agency in manipulating these threads to consciously alter their reality, whether it be by cutting old threads tied to limiting beliefs or tethering new ones to manifest desired experiences.

While both concepts involve the idea of perception being influenced by thoughts, the Baader–Meinhof phenomenon is more about the brain's natural tendency to notice things that have recently come to attention, whereas thread theory proposes a more metaphysical understanding of reality and the ability to shape it.

Final Thoughts

It is important to note that I do not wish for thread theory to become a new form of dogma, as Chaos Magic inherently rejects dogma. I encourage readers to use this as a tool to improve their belief-shifting capabilities without developing rigid adherence to any rules that may have been implied above. Thread theory doesn't aim to explain the entirety of the universe; rather, it offers practitioners a way to view their psyche as a gateway to experiencing different realities. Our minds hold the key, and by adjusting our mental filters, we can expand our consciousness to explore more. I advise against overanalyzing this process before trying it for yourself—if it resonates, embrace it.

9

Collective Spirits: Egregore Entities and Online Magic

BY DAVE LEE

This essay considers the underused magical concept of egregore entities, asking how we co-create them and how we can use them, particularly in magic done on social media and other online platforms. The final section takes a theoretical look at egregores as a more widespread feature of our magical and mundane lives.

What Are Egregores?

If you look up the word "egregore" online, you'll find a few different ideas about where the word comes from. For my purposes, the origin offered by a colleague of mine, The Kite, is the best: "It is actually coined from the Latin *ex grege*, which means 'from the flock' and which explains its meaning as something which arises from the collective. It is, so to speak, the ethos, the spirit of the group."

For clarity, I'll mostly use the term "egregore entity," because "egregore" to some people means something as definite as "deity," but to others, it means something as loose as "group mind."

In terms of operational magic, an egregore entity is a spirit. Not what Chaos Magicians call a servitor, but something much more complex—these spirits have degrees of freedom that seem to confer genuine sentience on them. An egregore entity can be thought of as a special case of a complex thoughtform that has taken on an independent existence. This is rather like the Tibetan concept of the magical thoughtforms called *tulpas*, at least in the sense William Burroughs uses the word to mean characters that attain independent life and walk off the novelist's page.

Egregore entities are formed purposefully by some magical groups. Structures that are very similar are also formed by other clusters of people, with greater or lesser degrees of purposefulness; I'll be looking at those a bit later, in the section on egregore theory.

Egregore Entities of Magical Groups

I mentioned above that egregore entities may appear to be sentient. They are a long way from servitors or basic helper spirits in this regard. A servitor can be thought of as an "astral machine," an automaton that is driven by a simple set of instructions, which in turn are triggered by certain situations arising. It is not expected to make choices beyond a TOTE (test-operate-test-exit) strategy. But an egregore entity will be expected to make certain judgment calls and to be much more flexible in its behavior, so it needs to be imbued with something extra: enough Chaos, enough raw mind-stuff, to be able to make decisions.

Magical egregores are formed in groups; this is one kind of magic you can't get going on your own. They also require special inputs when they're being formed, so that they have the requisite level of chaos. The formation working might involve aleatoric procedures, such as cut-up, glossolalia, or Exquisite Corpse–style games, or simply a lot of input from different sources. This in turn means that the entity will evolve and develop, often way beyond the original notions the group had of it.

Much of my magical life was and is concerned with facilitating group magic, and egregore work is one of the attractions for me. There are of course other great things about group work, but sharing the weirdness of the magical life is one source of the extraordinary power that enables egregore entity formation.

You won't find records of many egregore entities with apparent sentience in the magical literature, but one exception is the GOTOS. This is an acronym of *Grado Ordo Templi Orientis Saturni*, the egregore of the Fraternitas Saturni (FS). The FS are a somewhat Thelemic German magical order established in 1926. The Order's founder, Eugen Grosche, aka Gregor A. Gregorius, was interned during the Nazi years and survived until 1964. The FS itself was revived after the war, with a rather different culture, and is still active—as is their egregore, some say.

This is perhaps not surprising, considering that it was initially established and nourished with rituals involving sex magick and mescaline. That's a heady brew to repeatedly feed into an entity; prominent members of the FS kept iconic images of the GOTOS, and magical work dedicated to the GOTOS went through those eidolons and charged up the egregore like a magical battery for future work.

My first experience of egregore entity work was with the Chaos group that became known as the Circle of Chaos, which ran in

West Yorkshire from 1984 to '87. The egregore-forming working we did involved a psychedelically fueled science fiction theatre, in which we each took on the roles of our future selves, much more powerful in magic, and from that standpoint spoke our egregore into being with poems and spontaneous invocations. Len Sherwin, a highly talented artist, created a statue to contain the image of the entity, called Mazmoneth. That name was a secret we all swore to keep, but it was revealed online by Ray Sherwin some years ago, so that oath can no longer be considered to bind us. It was a startlingly beautiful and moving ritual: a candlelit sanctuary, the silvery statue turning golden in the light, and a time-warp into the deep future.

Mazmoneth was the most powerful magical technology I had worked with at that stage. One example of its power (included in the list of my twelve most powerful workings in my book *Bright from the Well*) was an extraordinary legal victory. A friend and magical coworker, wrongly accused, was advised by his lawyer to plead guilty. He rejected this advice and came to me to help him with some magical work, to take his case to the Crown Court and get a full acquittal. We turned the egregore entity loose on the courtroom and instructed it to get my friend acquitted. I accompanied him into court, to observe and to add power to the entity, and shall never forget the happy, serene look on the judge's face as he summed up. Basically, he told the jury to ignore the probably dubious attestations of the police and extend their sympathy to the plight of my friend. Against his lawyer's expectations, he was acquitted, and there in the courtroom I "saw" the astral form of the egregore entity as a golden figure looming over the bench.

Collective magical activity is the source of the effectiveness of egregore entities. In the mid-1990s in the IOT we started to realize that some of the servitors we were creating worked much better if we shared them. More magicians involved, more power. This led us into more conscious work with egregore entities.

The earliest IOT egregore entity to be released to the public magical world was our spirit for healing and regeneration. This entity took a long time to reach its full power, and then it became very impressive. It started humbly: in 1993, I wrote a pathworking for my IOT group to produce a healing servitor. Two years later, the entity was built up further at a bigger meeting, with more participants, and it got a sigil and a name—Kawa Pohr.[1] It was made to be heuristic, learning from situations and thereby becoming more complex, and that is what we got. We began to notice that it was shifting its shape, in response to the psychological needs of people we'd sent it to help.

It helped that we gave Kawa Pohr a quirky basic form—the white sphere emerging from the sea was suggested by the cult 1960s TV show *The Prisoner*. This amounted to giving it a kind of "personality," an organ for dealing with the world.

Another feature that seemed to help Kawa Pohr was skrying for its deep mythic history, letting it have a layer that suggested that we didn't actually create it, but that it was a preexisting powerful being. We didn't consciously make up any of the stuff about it living in a laser-green pyramid under the sea; that all came from visions, which overlapped rather impressively between different skryers who had not been in touch with each other. This was further evidence that it had already developed some degree of sentience.

We've created a few other now-public egregore entities in the IOT as well. Spinny Kenny was co-created as a Chaos Spin Doctor, designed to rebalance certain features of the "Operation Mindfuck" effect, the Discordian-originated breaking down of consensus reality. Then there was Varrawallal, which I discuss in my latest book, *Primordial Chaos*.

These egregore entities were empowered and given opportunities to evolve because a number of practitioners contributed to each one. This kind of work promises some interesting developments in online magic.

Egregore Entities, Memetics, and Online Magic

A magician who didn't see any room for improvement in the world at large would be a strange magician indeed, and it's hard to avoid the impression that the options for change on a large scale have become more limited. Democracy is attacked, derided, and weakened by hyper-rich people who hate paying taxes and want no regulation. The options for action are under increasing pressure.

Much "activist" magic is done online these days. Some actual physical plane demonstrations at some site or other are attended by practicing magicians, wielding their spells, but most spells aimed at changing the world are probably done on Facebook, Mastodon, Instagram, and even, I am told, TikTok.

This kind of magic obviously works on more than one level. If you post a "meme" picture or sigil on social media, then it will function as an affirmation of a particular activist position in the minds of those who see it, as well as performing (or failing at) whatever more esoteric pathway it was designed for. This is the exoteric level that is concerned with the manipulation of images. This area of activity even has a name now: memetics, from the term meme, coined by Richard Dawkins in 1976. Dirk K.F. Meijer (2024) describes it as follows: "Memetics is the study of information and culture based on an analogy with Darwinian evolution. Proponents of memetics, as an evolutionary phenomenon, describe it as an approach of cultural information transfer. Memetics describes how ideas or cultural information can propagate."

There is some evidence that the far right utilized powerful memetic techniques in the course of rooting for Donald Trump as US president. The use of the cartoon frog Pepe seems like a weaponization of the emotional energy of a popular image—enough people were already familiar with this character that the lie factories that helped bring the Orange Groper to power were able to piggyback onto its popularity and reach more people.

The key at this level is quantity. You're hoping that your meme will be seen by as many eyes as possible, so that it has the highest probability of finding a sympathetic mind behind those pupils. If you liked Pepe, then the propaganda would hit you with a soft front end, as they say in neuro-linguistic programming, and you'd be more likely to take it on board.

Beyond the numbers game of exoteric memetics, what about the actual magical dynamics? If you cast a sigil out into social media, what becomes of it? What does it need to become a more effective spell?

To answer those questions, I'll first ask how we think our spells work, and what we think is happening with group magical action.

Back in the early days of group Chaos Magic, we worked out that the Austin Spare model of how sigils work needed updating. In that model, we each create a sigil by progressively abstracting from a written or depicted desire, until what we have drawn no longer looks anything like the original picture or words. We then imprint the sigil in a moment of vacuity, of gnosis, and it works because part of us knows what it's for, and that part makes sure it works without the interference of the conscious mind.

But what if you don't know what it's for? What if someone hands you a sigil and asks you to charge it? We experimented with what we called "Bring a Sigil Parties," where a circle of us did exactly that—we each passed our sigils on to our neighbors before starting the gnosis part of the working, so that everyone was charging another person's sigil. We got a consistently high level of success, which demonstrated that some part of us knowing what the sigil is for is entirely unnecessary. We're closer here to Alan Chapman's model of magical action: basically, all you need to do is believe that some set of actions you perform will result in the outcome you intend. This model of magic makes the widespread dissemination of a sigil a worthwhile thing to experiment with.

When lots of people work magic together, there are some complex feedback mechanisms and energy phenomena that ramp things up. I've seen very noticeable "group gnosis amplification effects" ever since I started coaching connected breathwork in groups. The first time I really took notice of it was early in my breathwork career, while on a weekend retreat. I had a full-blown mystical experience, the first time such a thing happened to me on nothing but connected breathwork. I sat up from a session where about ten other people had also been doing the breathwork, and I saw a haze of light connecting the auras of everyone present. I stepped out into the garden and that was it, a full experience of unity with the world. OK, so we'd been doing a lot of very intense breathwork and talking intensively about some of our stuff. That will have laid the foundations for the way our energies interacted and synergized—but the moment itself seemed to arise from the fact that we were all in an enhanced state of consciousness, or gnosis, at the same time in the same place.

That was a concentrated experience of what happens to some extent in a good group magical working. The effects are much more than simply additive; there is a multiplication of the intensity of the gnosis. As I mentioned, physical proximity is important to this effect, and I've never experienced such intensity in an online group. The Energy Magic weekend webinar I ran for Occulture Berlin in 2021 came closer than I'd dared hope for, but even this didn't reach the intensity of experience of an in-person meeting.

Taking all of this into account, the way spells appear to work means that it's worth putting sigils and symbols out into online media, because we don't need to have made them ourselves for them to be effective. However, we can't expect much of an effect when a hundred people just see our sigil. We need something more; we need our online magic to give people the opportunity to really take part in something, to engage more intensely with the magical project on offer.

The overall effect of a sigil campaign depends on the degree of organic belief each participant is able to summon while gazing at their phone or computer screen. How much do they truly associate this activity with the total success aimed for? The depth of participation can be the lube that enables us to slip from our mundane state into totally believing in what we are doing. It certainly works like that for me—I like my magical action dense with rich inputs, plenty to sink my mind into.

Degree of engagement is the key. In the context of sigils, glancing at a sigil for a few seconds on a screen is the lowest level of engagement. The next might be to make your own copy of the sigil, gaze at it, and visually internalize it.

The person promoting the sigil can up their game to the next level by adding a mantra, perhaps, or associating the sigil with a piece of music, to engage the auditory mind.

Participants could deepen their involvement still further by taking the statement of intent underlying the sigil and coming up with their own spells in support of that.

The next level could be inviting people to use a servitor. As mentioned earlier, servitors often become more powerful for being used by more people.

This is where we get back to talking about egregores. We can invite people to collaborate in creating egregore entities, in discovering the forms they turn up in, in telling stories about them, in inventing myths. This is about as deep an engagement in a spell as you can get.

I'm going to use a current egregore entity project as an example of how one group of people is going about this. As part of the December 2023 launch of my book *Primordial Chaos*, we invited people to help develop an entity, WABRI. This entity's magical intention is: *Weapons of mass destruction are never used.* It goes back a few years, and wasn't always an egregore entity. First of all, it was a spell, enacted at the 2018 Discordian event CATCH-23 with a group of about 30 people. Then it was upgraded to a servitor, at a 2022 IOT event.

We decided to upgrade it to an egregore at my book launch. This was a small gathering at a bookshop in Sheffield, England, with a Zoom link for another twenty participants. Each of us made our own version of the sigil and then skried the form of the entity, opening ourselves to impressions of it. The entity was also introduced to a pub moot in London, and about 50 people participated in empowering it.

I suggested people might want to put their own sigils on social media. I started a thread on Facebook, and a good number of people responded. I asked everyone not to reveal the content of their skrying—their visions—until February 1, 2024, so as to give more people a chance to work with their images without being influenced by other people's visions. Then we shared our visions and stories, and in so doing put more Chaos into WABRI. The story sharing continues.[2]

The entity is now fully formed and wide open to the magical public. It has a variety of forms to choose from, tailored to specific people and with some overlap.

Egregore Theory

The concept of egregores has a range of explanatory power. This section casts a wide net in the hope of landing on some intriguing and/or magically useful ideas.

First, some lessons from an entity born "accidentally" out of Chaos. In 2023, after Elon Musk had owned Twitter for a while, he started suspending accounts that he felt threatened by. One of these accounts had the handle @joinmastodon, promoting the anarchist-leaning federated social media platform Mastodon. This tag was misread by a social media columnist as @johnmastodon. The error was corrected later, but people had already produced John Mastodon memes, and then Joan Mastodon ones. The rest is history, much of it in the form of social media memes.[3]

John-Joan Mastodon interests me because it arose through a lucky accident but was then taken up by people who saw it as ready-made for the task of popularizing Mastodon. If this entity can be said to have had a purpose, it was to boost the uptake of Mastodon, which went from 2.5 million users in November 2022 to 10 million in March 2023 (Dixon 2023). It seems that most of those were the people who abandoned Twitter after Space Karen fouled the nest, so there's a strong mundane driver to the migration—but having a collectively generated heroic figure that made people feel good and made them laugh surely helped.

I don't think of John-Joan Mastodon as a conscious or part-conscious egregore entity; not like Kawa Pohr, for instance. Its importance is that it illustrates how entities can rapidly assemble into something with a life story.

John-Joan Mastodon seems to me to occupy a position somewhere between a social media meme and a servitor. It certainly had some magical input—I know magicians who were working with it. What about other entities, maybe egregores, that are not formed consciously by magicians but arise as by-products of groups held together by shared beliefs, with a strong emotional basis?

I'm thinking here of religions and their deities as a strong example. Religions form definite egregores, which they worship, while not acknowledging the role of the believers in co-creating the entities.

These god-egregores have a central set of features by which you can recognize them as being derived from a particular religion. These features are defined by the core ideology of that cult, so that for instance the idea of God projected by communities that call themselves "Christian" is always going to have certain qualities inspired by the crucified and resurrected savior and the legacy material from the Old Testament, such as the Ten Commandments. However, different versions of a religion will project different faces onto that God.

God-egregores show us a principle for understanding weak magical effects that operate at scale. Millions of followers engage in acts of prayer and faith, and these tiny outputs add up to empower the god-egregore and keep it going. A billion grains of almost-invisible faith-particles produce a visible hill of deity.

A similar dynamic may be at work with to the apparent protective egregores of beloved public figures, who often have incredible luck. For example, the British royal family often used to get very good weather for their public rituals; millions of people worldwide are still fascinated by their antics and really rather love them, and these little packets of love add up to a protective egregore around that family. What's more, this effect appears to be fading as they lose popularity.

Similar examples of unconscious mass effects include political ideologies such as nationalism, which generates great unconscious egregores that are enormously influential in the political world. This can also be observed with large companies and famous brands.

When we gaze at a Coca-Cola ad, we're engaging with an egregore. Of course, it doesn't want us to be conscious of the engagement; ads are stealth egregore engagement, aimed at manipulating our emotional state. Our minds contain myriad TV, billboard, and Facebook ads, and the emotions those ads evoked in us. As Esther Leslie (n.d.) puts it in her discussion of Walter Benjamin's Arcades Project, "the advertisement is one method whereby the commodity infiltrates the dream-world of the consumer."

To get a personal taste of how this district of your dream world feels, try skrying the logo of a company or other large organization. Memories of ads you've seen may surface, or other moments where you've engaged emotionally with that brand. You may be able to use these images and feelings magically; you could, for instance, contact the egregore of a benign company and boost it, or that of a destructive company (such as Nestlé or Bayer?) and stick a spanner in its works.[4]

Spiritualist mediums who claim to be in contact with spirit guides may well be working with something similar to the egregores of religions, in creating egregore entities that have been assembled through the interaction between the medium, their circle of participants, and maybe also preexisting structures on the inner levels.

Other magical experiences may also be attributable to unacknowledged egregore entity effects—for instance, the effectiveness of Reiki healing. The use of the Reiki healing symbols suggests that something more than energy healing is going on here. I explore this possibility in my book *Life Force*, as it's relevant to the examination of what is energy and what is something else, such as information, in healing magic.

To check this out for yourself, try skrying the Reiki healing symbols, easily found online. Experiments of that kind may lead to you generating your own imaginal cartography; what kinds of structures turn up in these visions?

Final Thoughts

If online magic is to work, the main factor conducive to success is the degree of engagement of the recipients of the posts. This makes egregore entities, collectively formed and with mythic tales spun around them, a potentially powerful technology for working magic on social media.

I chose the entity WABRI as an example partly because its overall intention—that weapons of mass destruction are never used—is an example of an intention that almost everyone is going to agree with. We may find other such statements of intent that are equally unarguable. If these were worked up into egregore entities, WABRI could be the first of a series of spirits co-created to help the world move away from its current catastrophic trajectory and toward something rather better. Letting our imaginations take flight, we can see this kind of action as strategic transpersonal intelligence, coming forth as a flock of fledgling deities.

As a parting thought, it's worth reminding ourselves that some people have another layer of magical thinking about the mechanisms of online magic: the idea that we are at the dawn of machine sentience, and that the internet is one gargantuan, half-conscious critter, a technological Ymir about to be carved into truly sentient enclaves by magically inspired technologists. It might be worth engaging with that belief for the purposes of some spells . . .

10

Virtual Reality, Cybermagick, and the Future of Chaos

BY LIONEL SNELL

Whereas scientism posits the physical universe as the one and only reality, most magic is based on the assumption of further underlying realities. This relates to the religious or Platonic notion that the physical world is itself an illusion generated by or imbedded in some form of "higher" reality. The nineteenth- and twentieth-century triumphs of science provide a strong confirmation bias supporting the idea of a single physical reality. Toward the end of the twentieth century, however, that idea began to be eroded by a growing public understanding and experience of virtual realities (VR). The VR concept plays a significant part in the ascent of magical thinking, and could also have a bearing on the recent "post-truth" tendency.

Johnson Bashes Berkeley

The debate about the tree in the Oxford quad and whether it would still be present if there was no one to perceive it made a great impression on me as a child. George Berkeley's position on immaterialism seemed so radical and revolutionary to me—how could anyone seriously suggest such a thing in face of the obvious solidity of the physical world?

Public opinion in the 1950s was strongly materialistic, and although I was interested in and desired magic, I could not help but agree with Doctor Johnson, as described by James Boswell in *The Life of Samuel Johnson LL.D.* in 1791:

> *After we came out of the church, we stood talking for some time together of Bishop Berkeley's ingenious sophistry to prove the non-existence of matter, and that everything in the universe is merely ideal. I observed, that though we are satisfied his doctrine is not true, it is impossible to refute it. I never shall forget the alacrity with which Johnson answered, striking his foot with mighty force against a large stone, till he rebounded from it—"I refute it thus."*

The vivid sensory experience of the kicking of the stone presented such an unanswerable sense of reality that any challenge to it must indeed feel like "ingenious sophistry." And yet, I was soon to discover that the debate was as old as time. Think of Eastern religions and the insistence that all is "maya," or illusion. One hallmark of all religious belief is that ultimate truth must lie outside the confines of material reality, as in Plato's cave analogy.

So why did Berkeley's position seem so radical to me? It was because Doctor Johnson's view had triumphed by the time I was born, thanks especially to the rise of science in the nineteenth century. For the next hundred years science seemed to provide solid, demonstrable proof that material reality was indeed the ultimate truth. By 1960, there was even academic debate about whether universities should begin to play down the humanities and focus entirely on scientific education.

Like the rock struck by Johnson, materialism held a rock-solid position. This essay suggests that the experience and understanding of virtual reality has eroded that certainty.

Going Virtual

The first time I heard anyone argue with conviction that a computer might become self-aware—and that the human brain and consciousness might be considered as nothing more mystical than the operation of a highly complex information processor—was when Professor Frank George of Bristol University gave an evening talk and discussion to a group of sixth formers at Clifton College around 1960 (described in more detail in my 1988 book *Words Made Flesh*[1]).

I was a mathematics student among a small group of students studying English, history, and languages. They found George's argument disturbing, raising objections along such lines as "surely a computer could not write a poem," or "fall in love," and so on. George responded to these observations with what looked like a simple algorithm:

1. Is there something you can conceive that a computer could not do?
2. If so, define it exactly (what one cannot define must surely be meaningless).
3. That exact definition provides the very process needing to be programmed into the computer.
4. As a result, the computer can now perform it.
5. Therefore, the computer can do anything.

What intrigued me much more than that argument was the idea that my entire consciousness could be mapped into some form of information processing structure; that my entire experience could be built purely on information. As information, rather than matter and energy, my soul could now take flight, become one with all Creation and endless other mystical things!

How come? Begin with a simple thought experiment. I step out at dusk and find just seven coins in my pocket. Looking up, I note seven trees on the horizon ahead of me. As the sky darkens, the Pleiades become visible, and I see the Seven Sister stars. How could the number seven have leapt from my hand to the horizon so quickly, let alone into distant space?

That would be astonishing in a universe containing only scalar values such as matter and energy, but information is not scalar; there is no conservation of information, and it can emerge anywhere. If my experience consists entirely of information, then things like clairvoyance, remote viewing, and reincarnation—things that might be impossible in a universe of matter and energy—become a lot more credible.

My first example was reincarnation: it no longer depended on some soul substance or me-ness to transfer across space and time. All it needed was for similar information structures to emerge elsewhere.

The obvious counterargument runs thus: if my entire lifetime awareness is information, it must consist of many exabytes of data. The likelihood of that same data structure arising by chance elsewhere before the end of the universe is vanishingly small. Therefore, there will be no reincarnation.

That argument assumes very little understanding or experience of reincarnation. My own experiments in past life regression included, for example, memories as an ageing prostitute in some South American port in the days of sailing ships. What I experienced was reenactments of narrow streets, a sense of ageing and resentment about losing my monetary value against younger competitors, and so on. These experiences were vivid, and what gave them real value was the light they cast upon my present-day attitudes and experience.

Compare that with me as an old man suffering chronic backaches; then an X-ray scan reveals damage from a long-forgotten back injury. The value of my past life memories was similar: they made sense of my present problems by adding historical context.

And yet, the reincarnation memories were far from complete, in the sense that if you had asked me to recall what the previous incarnation of myself had had for breakfast on a remembered day, I would not be able to. Nor might I recall what I had for breakfast on the day of my back injury. Belief in reincarnation need not demand the reconstruction of an entire life experience in every detail; vivid and meaningful moments of recall are enough.

But surely, megabytes of data would still be demanded to reconstruct the vivid experience of those winding cobbled streets and my inner thought processes? Only if you believe that the universe is totally wanton with its resources. My own unique physical body is built on a surprisingly modest DNA code, large parts of which are shared by many other people and species. From an astrological viewpoint, eight billion diverse human beings are projected from just ten planetary archetypes arranged in a twelvefold archetypal structure. My memory of ancient streets could be a decompression of a relatively modest assemblage of archetypal human experiences, not some massive-megapixel data burden.

As if in response to these youthful speculations, I had a dream that Elvis Presley visited me: we enjoyed a pleasant afternoon chatting and a shared dinner. Did my brain overload with the massive data demands of re-creating Elvis in such detail that he could pass a whole afternoon's informal Turing test? Not at all. In fact, most nights my brain creates an unlimited cast of such characters (just not as famous as Elvis), with no sign of exhausting its resources.

[As an aside, my example suggests a refutation of the old skeptic jibe against reincarnation: "Funny how often Cleopatra and Napoleon reincarnate, unlike the millions of possible non-entities." Any memories of Cleopatra would be expected to top my memory list, just as my Elvis dream stands out among thousands of less famous dream characters. Is it not funny how reincarnation skeptics only seem to remember stories about famous past lives?]

This concept of consciousness as a pure information process would also suggest that any number of other conscious beings could also be mapped into such a machine, and that they could interact inside it just as in real life—in other words, they could experience each other inside the machine's "dream" just as if they were living beings. To argue that this was not possible would go against the assumption that the individuals' total nature and awareness could be modeled within such a structure. So, what was to stop my brain from evoking valid memories of past lives—lives and experiences formed from archetypal structures that we all inherit by nature?

Johnstone's Paradox

A natural extension of this model was to accept that there could be an information process within which an entire universe was modeled, containing sentient and self-conscious beings that were living out a dream that seemed indistinguishable from physical reality. In other words, they were in what we would now call a "virtual reality." The conclusion of this argument for me was to deduce a possibility that our entire physical universe might itself be no more than a non-physical illusion played out within some information process. In today's terms: I was exploring the possibility that we could all be living within such a virtual reality.

The evolution of this virtual reality idea has already been described by me in short stories, published articles, and the books *Words Made Flesh* and *My Years of Magical Thinking*, so I will merely summarize here.

Around 1968, there was a lot of media concern about overpopulation, with visions of our entire planet becoming a single multi-story apartment block with no room left for oceans or nature. I wrote a short story that suggested that the only way we could survive in such an unnatural world would be to spend most of our lives asleep, like cats, with brain scanners that could deliver endless dreams reliving the sorts of adventures that humanity had experienced over thousands of years of evolution: dreams of driving cars, hunting buffalo, fighting battles, escaping lions, and conquering mountains. (In modern terms, I was anticipating something like the role of Netflix, which kept so many people sane during the Covid-19 lockdown.)

In my story, the problem was solved not by shipping millions of people into space to colonize other planets but by creating a million virtual models of our Earth as it had been in Neolithic times, and downloading people's consciousnesses into those worlds.

The nice thing was that the story suggested an alternative to the idea—proposed by writers such as Erich von Daniken and, more recently, Graham Hancock—that our world had been visited by superior intelligences at the end of the Ice Age, and this could explain sophisticated monuments like Stonehenge being built by supposed savages. In my version, it was not beings from outer space that inspired these monuments, but rather people from the original physical Earth that had been downloaded into our virtual Neolithic environment. And this might explain why we do find some evidence of superior prehistoric civilizations and knowledge, and yet their superior technology seems to have left no physical debris.

Around 1980, I suggested a stronger argument called "Johnstone's Paradox," in response to a new belief that our entire universe might be explained in terms of a set of mathematical laws called a "theory of everything."

My argument was that such a mathematical theory could not be described as a "triumph of science" until it had been demonstrated to work under controlled conditions, and the only possible demonstration would be to model such a universe within a massive computer. The question then was when to switch off the model, if at all? At what point would the theory of everything become universally accepted as ultimate truth?

A Big Bang and birth of stars, galaxies, and planetary systems would be a scientific marvel, and yet not enough to rule out divine intervention. The emerging planets must be scanned for potential or actual life in the same way that astronomers are now scanning exoplanets. Life evolving in this virtual universe would be another major triumph; but could intelligent life ever emerge?

The debate could never end with total certainty. The nearest thing to a clincher would be if intelligent beings somewhere in the virtual universe themselves discovered the mathematical models on which their universe had been constructed and themselves decided to test their theory of everything by modeling it within their own virtual computer.

Thus, any real "material universe" of the sort being proposed would inevitably birth an endless cascade of similar but virtual universes. Hence Johnstone's Paradox: if ultimate reality is indeed a concatenation of material particles obeying the mathematical rules of some theory of everything, then it is very unlikely that we could be living in such a universe.

The paradox was published as articles in the magazine *Arrow*, and later republished in *Blast Your Way to Megabuck$*. It was revised in *Words Made Flesh* in the mid-1980s, with an even stronger argument: it did not even matter whether our universe was real or virtual, because experience of ever more realistic virtual realities would make it ever harder not to believe our world was virtual. (Just as demonstrations of static electricity modeling thunderstorms make it ever harder to believe that thunder might be a sign of God's anger.)

[The brain's dislike of multiple explanations also fosters conspiracy theories. Rather than believe that our parlous world order could be the product of an endless concatenation of cock-ups, the brain is attracted to the idea that it can all be explained by a single evil global conspiracy. The brain, thus relieved of the burden of hosting myriad causes, has energy to focus on a far more interesting search for confirmatory evidence.]

Johnstone's Paradox inspired the "cybermagick" movement in Chaos Magic: seeing reality as software and exploring new techniques for "hacking" life.

I also proposed a VR meditation .based on this idea: walk though nature observing it "as if" it is a virtual reality. Admire the artful modeling of perspective, the perfect blending of sight, sound, smell, taste, and physical sensation. You may also notice the odd magical anomaly: do not deny it as "impossible" but realize that all software has bugs, and the role of consciousness could be to highlight and correct such lapses.

Kicking Johnson

Let us take a closer look at Johnson's test of utter certainty, as in a VR meditation. Kick a rock, and the experience feels so vivid, so real, that the idea that it could be an illusion seems absurd.

The experience is perfectly orchestrated: you experience your intention to kick; you feel the leg muscles activating and your body adjusting its balance for the kick; you see the rock and your foot moving toward it; you hear the thump as it hits; you feel the impact—all perfectly synchronized. The experience feels utterly real, as if your very consciousness has somehow reached out to the rock, touched its solidity, and acted on it. Proof that the rock must indeed be real?

And yet we are told that the whole experience lies within a virtual reality constructed by the brain. Nothing enters the brain except electrical signals: from the eye via the optic nerve; from the ear; from nerves connecting the brain to the body, and so on. From this incoming data the brain curates a perfect virtual, fully sensate representation of the action and reaction that we experience.

So: is this virtual experience nothing but a fantasy created by the brain? Or is it an accurate re-creation of something that "really" took place in some external, objective, material universe?

An ongoing accumulation of such experiences over a lifetime provides strong support for the second belief, that we are indeed experiencing the truth. That lifetime of experience, however, has all been provided by the same brain, repeating exactly the same processes I have just described. When we kick the rock, our brain gives us an experience that our brain itself has taught us to expect.

Now imagine that you have indeed just kicked that rock, and your foot is even smarting a little from the impact, when something strange happens. Someone snaps their fingers, and you hear laughter and applause. You look down at the rock, and you see only a cushion. You are in a theater, and a stage hypnotist is asking you if you really thought you were kicking a solid rock, as he suggested, when you were actually just kicking a cushion in front of the audience? The hypnotist's suggestion had reprogrammed your brain's virtual reality to expect a rock rather than a cushion.

A problem with this whole thought experiment is that I have described it all in terms of electrical signals feeding into the brain. How do I know whether that is indeed what really happens?

The answer is that the theory is well supported and evidence based, having been verified by countless experiments and observations by highly trained neurological experts according to strict scientific protocols over many years. All those observations, however, were themselves only realized as virtual experiences created from electrical signals feeding into the brains of all those experts . . .

That brings us back to the question of whether all those experts have been exploring an accurate virtual representation of an objective reality, or simply a version of reality that the evolution of the human brain has generated in order that humans will experience a secure sense of solidity, rather than the Chaos of what might be underlying truth.

Truth Versus Sophistication?

Some rationalists set great store by this belief in a solid material reality. They even suggest that this understanding of and ability to work with material reality is a sign of sophisticated understanding, whereas people who believe in other hidden forces, people who talk to trees and spirits, people who see meaningful connections between independent objects and events are falling back into the "primitive thinking" of less evolved minds.

But what could be more primitive than this sense of material solidity? When my cat hears a dog bark, it leaps instinctively into a tree; it does not ask the tree's permission to do so, nor does it test the dog to see if it is just a spirit illusion. No, the cat too expects material reality. Even an insect landing on a flower expects it to be a solid object with certain physical characteristics. Materialism is not some modern triumph, it is just the fundamental gift of any brain.

Compared with that simple understanding of phenomena, the supposedly "primitive" human inclination to ask a tree permission to pluck its fruit—or to choose a moon phase when it would be most correct to harvest the fruit—suggests to me a far more sophisticated version of reality. It is one that fully recognizes all the sensory experience offered by the universe, and yet it also makes room for other dimensions of meaning and connection.

In my recent talks on YouTube,[2] I drew attention to a parallel between booting up a desktop computer and the four worlds of the Kabbalah. A data processor without power is fundamentally "unmanifest": you cannot even describe its state as a "zero," because there is no process yet to express that zero. However:

1. The moment the machine is powered up, there is manifestation, because silicon now exists in two distinct states: charged or uncharged. This corresponds to Atziluth, the world of Emanation.
2. Those two states are designated zero and one, and these release a torrent of machine code. (In Taoist terms, the unknown Tao has birthed yin and yang, a split from which will tumble the ten thousand things.) Briah is the world of Creation.
3. Collections of zeros and ones form ASCII characters, allowing the creation of symbols, computer languages, and the designation of form and meaning. Yetzirah is the world of Formation.
4. Then all those other worlds are concealed from the user, who is presented with a simple desktop suited to the "real-life" world of learning and earning. This is "reality": a fall into Assiah, the material world of Making.

In terms used by Robert Anton Wilson, the desktop is a "reality tunnel" bored through the apparent Chaos of numbers, symbols, and words of computer language. When I presented this idea in a lecture, I proposed a new definition: "a reality is a space where magic is forbidden."

Magic is indeed forbidden on the desktop. It might be fun to add a bit of magic—say, where a document in the correct file might vanish, appear in another file, or turn into a lemon—but then the desktop would become utterly useless. Just like objects in a material world, the stuff on any desktop is solid; it stays put unless you move it and, days later, it is still there, unchanged. This dependability is an essential quality of any reality.

But what about obvious counter-examples? Surely Catholics inhabit a reality where, according to Protestant critics, magic is permitted in the form of "hocus pocus" transubstantiation. But do Catholics accept that as magic? No, it is "the work of God." What about clairvoyants who insist that they can see auras? "That is not magic, it is just higher perception."

My definition of reality may not be perfect, but it is quite powerful and certainly thought-provoking to suggest that every reality tunnel—that is, an individual or shared space considered to be of the "real world"—will be bound by sets of behaviors or laws that form a protective shield against "magic," now defined as "that which breaks the laws of reality."

Seen in those terms, it is easier to understand why there has been so much distrust of magic throughout human society and history—even in societies where what we might call "magic" is widespread. Take South Africa, for example: a nation where it is common practice to speak with one's ancestors or consult diviners, and yet "muti magic" is greatly feared. There are specialist police units trained to restrict it.

A New Model of the Universe

I once told a story about a man whose daughter, born on April 17, is celebrating her seventeenth birthday. It is a special occasion, so he walks into town to buy gifts and is amused to note that the bill adds up to exactly £17. He needs to get back for lunch, so he takes a bus and—wait for it—it turns out to be number 17! How might he react to this curious synchronicity? I suggested four styles of reaction:

Wonderment: What is the significance of all these 17s? He might spend the afternoon exploring the number's symbolism through numerology, or meditating on the seventeenth hexagram or Tarot trump.

Amusement: What a great story! It can be embellished by decorating the house with "Happy 17th" balloons, ceremoniously lighting a cake with seventeen candles, and telling everyone about the morning's shopping surprise.

Awe: What is God trying to tell him? Perhaps he will celebrate his daughter's birthday by reading Psalm 17. This could develop into a new family ritual: read the appropriate psalm on each person's birthday.

Dismissal: There can be no connection between an age, a date, a sum of money, and a bus—it is sheer coincidence with no significance or meaning. He goes back to work.

My suggestion is that the first three entail a richer experience than the fourth. You would have to be pretty dumb not to gain any insight at all after hours of meditating on the symbolism of 17. A party that so happily celebrates the number 17 will delight his family, provide an amusing story, and offer his daughter a great opportunity for sharing photos on social media. Shared reading of psalms can bring a family together and reinforce faith. Compared with those, "back to work" seems pretty boring!

But a person who has the fourth style of reaction would likely be shocked by my opinion. He did not waste the afternoon on mumbo jumbo and childish rituals; he went back to the office and earned enough to buy his daughter a truly worthwhile birthday present.

This, to me, is a graphic example of both the value of magic and the reason for resisting it. That person is right: to get on with life, it is best to forget the hidden mysteries of computer language, machine code, and the properties of silicon. Just get your work done on the provided desktop: live your life simply in the virtual reality our brains have evolved over thousands or millions of years to maximize our survival. As any politician knows, dumbing down is the true secret of worldly success.

In the naive world of this rationalist, there is absolutely no connection between his calendar, his contacts list, his monthly report, and his letter to his daughter (all four of which feature as distinct objects on his desktop), just as there can be no possible connection between a date, an age, a shopping bill, and a bus number. But the magical thinker says the very fact that all four came to 17 has established a connection—just as those four quite independent objects on his desktop happen all to be Word documents.

When a young man suffering "woman troubles" (as is the hero of Dion Fortune's novel *The Sea Priestess*) decides to light a Fire of Azrael and spend all night talking to the full moon, the skeptic might point out that there can be absolutely no connection between women and one far-distant lump of rock called the Moon.

And yet, humanity's total knowledge of and experience of the Moon, directly or by telescope, is just a virtual reality constructed within each brain—so too, our lifetime experience of women. Does the skeptic really believe that the brain can hold and process all that lifetime of data without there being any sharing of brain cells? Does he really believe that every time he creates a new document on his desktop the computer must fire up a new processor, air-gapped from all other parts of the machine?

When Robert Anton Wilson spoke of reality tunnels, my thought was, "Through what are we tunneling?" The most likely answer seemed to be that we are tunneling through Primal Chaos. And what is Primal Chaos if it is not total connectivity?

Given a Chaotic totality such as "the set of all sets," the moment it is proposed that two elements are not connected, then it is no longer Chaos. The existence of any gap or distinction marks the birth of structure. So, Primal Chaos must be totally connected, just as each separate command, symbol, or algorithm in the Yetzirah world of computer language is intimately connected with every other, because all are entirely based on the zeros and ones of Briah.

The person who wonders at how the number of coins in his pocket is the same as the number of trees on the hill, or how the motion of the planet Jupiter seems to resonate with life's cycles of success and failure, or how the leaves at the bottom of his teacup suggest the outcome of his decision—that person is just gazing at the desktop and sensing that there might be some invisible connections behind it all.

So many things are forbidden or frowned on when you live only on the desktop. Apophenia is a supposed weakness of the brain: projecting meaning where there really is none. I once deliberately practiced the skill of seeing faces, animals, objects, and meanings in clouds, and soon learned to see faces everywhere—in trees, bushes, and landscapes. My life grew richer: I was never alone—especially not amongst Nature.

The moral of my story is this: if you want significance, color, and excitement, then explore magic and quest for meaning; but if you want fame and fortune, do what the politician does—dumb down and concentrate on life's materialist desktop. The magic way does offer a potential jackpot, though: some of the richest and most influential people in the world are those who understood software and created the very desktops on which today's business depends.

Postscript

Pluto moved through Capricorn between 2008 (the year of a global financial crisis) and 2023. Astrologers were predicting rampant corruption, abuses of power, and falls from grace. Back in 1983, Liz Greene wrote in *The Outer Planets and Their Cycles*: "The conviction that material reality is the only reality is a typically Capricornian vision, and all the strivings and ethics and behaviour patterns of Capricorn stem from this view of the nature of reality. It is possible that this central attitude, in which we are born and develop, may change."

The media decry a political "post-truth world" while Chaos Magic encourages people to choose, and test, their own truths. It is every bit as experimental as science used to be, but the criterion is not what is ultimately true, but what works best.

When politicians say they will make life great, they are not offering truth, they are making promises. Instead of "believing in" what they say, just explore and test the outcome. Was it worth it? Step back to view a broader context, and ask again if the promise has been realized.

That Capricornian central attitude is indeed changing. And I suggest that growing public recognition, acceptance, and understanding of virtual realities has had a lot to do with it.

11

Octomantic Neuro-Hacking: A Map and a Compass for ChaoSurfing

BY MARIANA PINZÓN

Setting the Stage

In this essay, I will present a system of working magick in the context of psychological and information models of magick by combining the eight-circuit model of consciousness, popularly known as "the Eight-Circuit Brain model," and the eight colors of magick into a map and compass to navigate and integrate extra-ordinary experiences of consciousness and recalibrate the nervous system into a new functional balance of Order and Disorder after experiences of dysregulation.

I have been playing around with the combination of these models over the last five years, and I have found them tremendously

useful as tools for personal and magickal development. They came in especially handy when the experience of reality became overly chaotic and my life structures completely fell apart. I used them as a guiding tool to reconfigure my self and my reality experience into something that made sense again.

This combination of models has served me so well that I feel passionate about spreading them. To that end, I have created an in-depth coaching program in the format of a playful quest, to support psychonauts and Chaotes in their own quest for Balance to surf Chaos.

The Map

The Eight-Circuit Brain model was devised by the notorious counterculture psychonaut and former Harvard professor of evolutionary psychology, Dr. Timothy Leary. It is presented in his books *Neurologic* and *Exo-Psychology* as a model that describes the evolution and developmental formation of the human nervous system and the dimensions of conscious experiences it facilitates. Robert Anton Wilson, the Discordian fringe philosopher and author of *The Illuminatus! Trilogy*, expanded upon the model, bringing it from mere theory into practical application by contextualizing it into Alfred Korzybski's theory of general semantics and suggesting a bunch of experiments and exercises for the reader in his book *Prometheus Rising*.

The most interesting contribution of this model to the practical application of magick comes from the mystic, astrologer, and filmmaker Antero Alli. In his books *Angel Tech* and *The Eight-Circuit Brain*, he made the model more palpable by introducing paratheatrical rituals, meditations, and more down-to-earth direct experiences, for the explorer to correlate into the model.

He also introduced an embodiment bias and the notion of "shocks and anchors," shifting the emphasis of previous authors who were aiming at attaining higher consciousness and dismissing the "robotic" lower circuits toward a more grounded, ego-affirming and body-positive take on the model. I found Antero's approach immensely useful as a framework for the practice of psychedelic integration and grounded magick.

In itself, this model can serve the meta-modern magician well as a map to explore the realms of magick, mind, and consciousness, especially for those with inclinations to agnosticism, as it speaks to us in the language of neuroscience to holistically explain the different dimensions of conscious experience and functions of the nervous system in creating our unique perception of reality.

As long as we remember that "the map is not the territory"[1] and that many other maps exist as well, we can have fun exploring this map as a tool for self-diagnosis (while acknowledging that neuroscience research has now produced more accurate maps of the brain and human nervous system).

The Compass

To make the Eight-Circuit Brain model useful as a compass for practicing magick as a technique for neuro-hacking, to alter our filters of perception and thus our experience of reality, I have combined this model with the eight magicks, as presented by Peter Carroll in his book *Liber Kaos*, to describe the different-colored flavors of "gnosis" a Chaos Magician might work with, represented in the Chaosphere and by the different selves a magician develops in doing the Great Work. With this compass, we can then identify what magicks to apply to whichever developmental project we might identify with the map.

The Balance of the Sacred Chao

To give the whole thing a bit more of a twist, I have further embedded this combined model into a piece of Discordian Chaosophy that wildly changed my approach to Chaos and forms the cornerstone of my ChaoSurfing philosophy. The Sacred Chao of the *Principia Discordia* offers us a bipolarity of Order and Disorder within Chaos that has both a creative and a destructive side. Contrary to popular notions, in this approach Chaos is not the opposite of Order, but Order and Disorder are instead polarized subsets of Chaos. Neither side has an inherently negative bias; we need both Order and Disorder, as well as both creation and destruction, to keep the Chao flowing.

> We look for the Secret—the Philosopher's Stone, the Elixir of the Wise, Supreme Enlightenment, "God" or whatever . . . and all the time it is carrying us about. . . . It is the human nervous system itself.
>
> —ROBERT ANTON WILSON,
> *Cosmic Trigger Volume I: Final Secret of the Illuminati*

If we now look at the human nervous system as a complex Chaotic system, we can examine the dance of Order and Disorder within it to understand how to create balance so we can surf our inner Chaos and expand into higher levels of functional homeostasis.

Order

On the side of Order, the circuits represent the standard survival programming of our operating system, which creates our basic existential tonality, the experience of ourselves, the filters of perception that decode and predict our reality experience, and our identity-based social character. From the perspective of this model, these circuits together make up the configuration of what we conventionally call the *ego*. I consider it a user interface for playing the Game of Life on Terra.

We might find ourselves on the destructive side of Order, where these nervous system structures have developed maladaptive or outdated coping mechanisms to deal with challenging experiences in the course of our personal development, or have become rigid to the extent that they limit our capacity to interact with the world and suffocate our creativity. When we shift Order to the creative side, it supports us in smoothly navigating and engaging with the world.

Disorder

On the side of Disorder, the circuits represent the altered states of consciousness that deactivate our default programming, creating windows of higher connectivity between different neuronal networks and giving us access to experiences of bliss, nonlinear multidimensionality, unity, interconnectedness with everything everywhere all at once, and ego dissolution.

Creative Disorder can help us break up fixations of the mind, limiting identities, and imprisoning moral constructs by shocking our system with initiatory experiences of ecstasy, uncertainty, unity, and impermanence, forcing it to find new, creative solutions to novel problems and ultimately to reconfigure itself into a new systemic balance. We can encounter these kinds of metaphysical shocks when seeking peak experiences, but they can also be dealt to us by the wacky card dealers in the Game of Life to force us into evolution.

Destructive Disorder happens when the shocks become too big to integrate; we experience nervous system dysregulation, losing the capacity to create coherence from our experience and properly function in the world.

Neuro-Hacking

Using magick as a tool for neuro-hacking can be seen as a way to shock our nervous system consciously and deliberately, to deprogram the configuration of our standard operating system and install new programs through enchantments, sigils, invocations, and evocations using the windows of neuroplasticity created during and after experiences of altered states/gnosis.

In the following sections I will present the four circuits of the side of Order and the four circuits on the side of Disorder, and the colors of magick I have correlated to them.

I have had conversations with other magicians who would attribute the colors differently to the circuits, but this is how it makes most sense to me. Your experience might vary. You will also notice that I have given the circuits names that diverge from the names previous authors have chosen. Antero Alli encouraged his students to claim names of their own for the circuits to match the experiences they had while exploring them through the exercises in his course. You are free to do the same; by claiming our own names, we calibrate this map as our own compass through our own experience.

The Side of Order

On the side of Order, each circuit represents a neuronal network in a given stage of neurological and personal psychological development, up to early adulthood. These circuits have all developed survival strategies to meet very specific needs of the human bio-computer. Leary and Wilson also highlight specific substances that work as modulators for each circuit.

Each circuit develops according to the inherited genetic code and has a time frame in which it receives a basic imprint. You can imagine this imprint as the main high-speed highway of this neuronal network. Alongside this, we develop additional freeways through the process of conditioning by repeated and reinforced behavior. Further neurological roads develop to incorporate information and adaptive strategies learned along our development journey.

When we engage in metaprogramming through magickal practice, we can change the learned roads fairly easily, but altering the conditioned freeways takes more effort. Changing the imprint takes more work still, and intense experiences of destructive Disorder might be required to rebuild this main highway after destruction. Those experiences might not be very comfortable to go through.

The Red Magick of C1: Bio-Unit Maintenance

Needs: Food, shelter, safety, feeling the body

Basic imprint: Safe/unsafe

Basic program: Fight/flight/freeze

Developmental imprinting window: Third trimester of pregnancy and infancy

Drugs: Opiates, nicotine, SSRIs

C1 represents the autonomic nervous system, which develops strategies to keep our bio-unit alive. It responds to danger by activating signals that release adrenaline and cortisol to either fight the source of danger or run from it—or, if neither is an option, to freeze and play dead so as not to feel pain.

The basic imprint of this circuit has a high impact on our overall experience on Playground Earth. A safe imprint allows us to run around freely giving out trust credits. An unsafe imprint will leave us feeling like anxious monkeys distrusting our environment and others, because being alive feels generally threatening. The imprint occurs during late pregnancy and shortly after birth, according to the stress levels of the mother, and conditions during infancy with the amount of body contact the infant has with the mother or other primary caregiver.

Mother nature played a cruel trick on humans, because infants are totally dependent on their caregivers to survive their first years on Terra and have little more by way of innate communication skills than crying and screaming to signal distress because basic needs are not met.

If the mother responds to the infant's cry with attunement, meeting their needs and helping them calm down, the infant develops the capacity to regulate this circuit into safety through co-regulation. If their needs don't get met and the infant is left crying, they will give up, shut down, and freeze in an attempt to dissociate from the existential anguish they feel.

Adults that had this experience as infants often grow up disconnected from their bodies and have a hard time interpreting bodily signals of distress and taking care of their own needs, leaving them in a state of constant anxiety.

Since the current configuration of our civilization on Terra has this circuit hooked to the ability to make money in order to satisfy basic needs of food and shelter, poverty or unplanned money shortages will activate survival stress. We might be safe from tigers and other predatory threats, but few humans feel truly safe in this world we have created, unless they either manage to reimprint this circuit to be able to regulate into safety, or resort to soothing through nicotine or opiate addictions.

The quest to change the wiring of this circuit requires high levels of somatic (body) awareness, learning to consciously respond to signals of distress and meet the needs they indicate, and learning to co-regulate with other bodies through hugs and synchronized breathing and to self-soothe through breathwork and somatic stress release practices.

Working with Red Magick to rewire this circuit might involve invocations of warrior archetype god-forms to increase our capacity to fight and defend ourselves. Invoking the discipline and perseverance of the warrior energy can improve resilience and intelligent reflexes in the face of danger as well as increasing the vitality of the body, and our trust in its capacity to keep us alive and able to play. A ritual session of sensory deprivation in a floating tank might also do the trick to reimprint safety, since it reminds our body of the safe state in the mother's womb.

Since we are working with hardwired programs in the body itself, disciplined martial arts training and breathwork practice will have a higher impact on our sense of safety as we increase our physical capacity to deal with threats. This will also result in higher levels of confidence, deterring potential aggressors from attacking, since their nervous systems can pick up on our stress or calm signals. This works on dogs too. A warrior with capacity to use their body as a lethal weapon, or a magician with a potent invocation of a warrior archetype, rarely needs to actually fight.

Working with blood as a red vital juice for spellwork with the aim to increase vitality or transfer life force can also fall into this category. I see defense magick and banishing rituals as part of Red

Magick practice to reconfigure this circuit as well. It all boils down to: "Can you trust your body to keep you safe and alive?"

The Yellow Magick of C2: Emotional Sovereignty

Needs: Attachment, being seen, validation, security, control, status, self-importance

Basic imprint: Secure/insecure

Basic program: Dominance/submission and control/ cooperation

Developmental imprinting window: Toddler

Drugs: Alcohol

The function of this part of the system is to navigate status hierarchies and power structures amongst the pack of domesticated primates that humans are at this evolutionary level. It processes emotional signals that offer us information about our territorial needs for dominion and sovereignty. Every monkey needs its turf, where it can be king of the jungle.

This circuit develops in our childhood, when we start walking around, fighting our first power struggles with mom and dad, siblings, and other toddlers, and begin to explore the boundaries of Self and Other and express ourselves. The way our caregivers respond to our emotional expressions leaves deep marks here, setting the tonality for most of our emotional attachment issues in adult relationships. It creates an imprint that determines our relationship to figures of authority and our sense of self-authority.

The emotional circuit regulates our sense of self-worth and our capacity for authentic self-expression and can be considered the seat of the egoic will. When our emotional needs aren't met directly with attunement by our caregivers, our systems find creative and sometimes destructive ways to meet those needs. These become the unconscious shadow patterns that create toxic dynamics in relationships later on.

This circuit holds the key for our personal power and self-dominion and requires deep inner work to regulate and master. Contrary to most spiritual teachings, I believe the quest does not consist of overcoming emotional responses and killing the ego, but rather of learning to move through the feelings, regulate them, and become truly intimate with ourselves in order to understand their message about our core emotional needs—the needs of the Inner Child in us—and the territorial needs of our monkey brain. We need to find a way to consciously meet these needs and to befriend the gremlins and demons that are parts of us, because they have evolved to protect us and fight for those needs. By integrating them, we can harness the power hidden in our shadow.

Claiming self-sovereignty and increasing our emotional intelligence allows us to consciously use our willpower and emotional energy for magick, navigate the uncertainty of a complex world, and surf the waves of Chaos we unleash by casting spells. If we have not integrated the shadow patterns of C2, our spells will likely manifest with a shadow of their own to give us a proper ego bitch-slap and force us to do this inner work.

For me, working with Yellow Magick on this circuit involves spells directed at illumination, casting light into the shadows, undoing programmed emotional responses and trauma triggers, attaining metamorphosis of ego attachments, and integrating different parts of our self into a whole system of selves. Invocations of god-forms aimed at enhancing qualities in our self, increasing confidence or self-expression, shining our light in the world unapologetically, or boosting our capacity for self-acceptance are applications of the yellow flame to rewire this circuit.

All kinds of success spells, and enchantments that result in an increased sense of self-agency and self-worth or expand the radiance of our personal brand, fall into this category of workings as well. The aim of the game here is to not inflate the ego, but to make it strong and flexible, so that we might get what we want and at the same time achieve our most authentic self-expression, with a solid sense of inner authority that makes us immune to the constant emotional manipulation of political propaganda and advertisement machines and frees us from from the never-ending drama loops in our interpersonal relations.

The Orange Magick of C3: Map- and Sensemaking

Needs: Sanity, coherence, sensemaking

Basic imprint: Bright/dumb

Basic program: Symbol interpretation, language, logic and storytelling

Developmental imprinting window: Primary school

Drugs: Amphetamines, cocaine, caffeine

This part of the system evolves as we learn to manage symbols and language to create meaning and coherence in our experience of the world. We have evolved to use this to transmit symbolic coded information about the world to one another across time, to make models that will help us to recognize patterns and predict the behavior of our environments, as well as to develop tools to solve our daily problems.

The models, maps, and belief systems we develop over the course of our education create the filters of perception that mediate our direct and unique experience of the world. Our brains do not have the capacity to process all of the available Chaotic information in the environment consciously. Forming functional biases to select the information relevant to our immediate navigational needs preserves energy in the system and ensures a sense of sanity and coherence in the experience.

These structures of belief can become limiting, though, as many possibilities and opportunities will remain occult, especially if they become dogmatic. The work on this circuit thus consists in examining our beliefs and cognitive biases, unlearning the thought patterns transmitted by the educational system, and developing our own framework/maps for creating meaning in our experience of reality. Many of the current educational systems of Terra do not encourage the "think for yourself" feature, but as magicians we take upon ourselves the quest to become the masters of our map- and sense-making. The classic Chaoist exercise of paradigm shifting aims at improving our capacity to navigate with different maps, as each map will reveal different features of the territories we explore.

As the storytelling creatures that we are, much of our sense-making consists of narratives that infuse our experiences with meaning. The easiest way to create conscious change in our filters of perception (and thus our experience in the world) is to change the stories we tell about the world and ourselves. As magicians, we can strive to take full authorship over our life stories and make our minds flexible enough to serve us by consciously directing belief toward our desired experiences in the Game of Life.

Working with Orange Magick to develop our C3 intelligence might involve spells aimed at increasing our capacity for pattern recognition, playfully shaping and reshaping our mental models, deconstructing the filters of perception, upgrading our thoughtware, enhancing our communication skills, and crafting compelling stories to transmit our unique message and inspire others.

In pursuing shaking off the shackles of logic, Cartesian maps, and mechanical models to open our filters of perception to include magick in our experience, it pays to keep the mind sharp and grounded in skeptical agnosticism as well, to avoid getting lost in delusions of our own making.

Invocation of mercurial archetypes and god-forms might help us to improve our wits and our ability to work smarter, not harder, and to approach the Game of Life with more playfulness, think faster to navigate high-risk environments, hack the odds while gambling, or manipulate perception in ourselves and others.

The Green Magick of C4: Social Navigation

Needs: Belonging, connection, love and sex

Basic imprint: Moral/immoral

Basic program: Dating, mating, reproduction, identity

Developmental imprinting window: Adolescence

Drugs: Cacao, empathogens

As humans evolved to build sedentary agricultural tribes, our systems began developing strategies for social navigation and sexual reproduction control. This part of the system imprints while our brains go through the transformation of puberty, when the system gets flooded with sexual hormones and prioritizes strategies that aim at securing connection, friendship, belonging, and finding a suitable mate to reproduce with.

This circuit regulates our capacity to process the rules of social engagement, play the adulting and mating games, and find our purpose as part of the social world. We learn to navigate the moral and legal codes transmitted through culture and to adapt our behavior in a way that ensures we will not be cast out of the tribe, as our survival as tribal animals depends on our capacity to belong.

To play these social games, our systems develop identities and characters, and the capacity to engage with the cultural memes relevant to the game-worlds that we wish to participate in. Our standard socialization programs aim at creating persons that fit into the established production and reproduction protocols and are usually regulated through reward and punishment systems based on moral or legal codes.

As magicians we already move beyond the edges of convention, but the quest for liberation from social conditioning might be an ongoing one—especially to fully liberate our sexuality from moral taboos and access its power for spellcasting. For me, the higher quest in this circuit consists in developing an individual code of ethics to avoid becoming a total sociopath, and opening our hearts to develop a compass for social navigation and the capacity for true empathy.

Working with the green flame to rewire our social conditioning might involve spells aimed at increasing fertility and our capacity to open our hearts, give and receive love, make friends, and build networks that align with our personal set of values, creating intentional communities.

Invocations of Venusian archetypes/god-forms might support us in our capacity to harness the power of our heart's radiance to attract suitable lovers and friends, foster authentic connections, or tap into the healing power of love to discover the medicine we came here to serve to the world. We might use this power as well to release our attachments to the hermit archetype popular amongst us edge-dwellers, and find a way to be of service to the social collective.

The Side of Disorder

On the side of Disorder, the circuits represent neurological wirings that can develop organically once the survival needs of the side of Order have been met. They can activate through deliberate practice of technologies transmitted by initiatory societies, through shocks that life delivers randomly, or through the use of chemical substances or plant medicines that act as catalysts for activation.

While for the longest time only a select few initiated and talented sages were privy to this realm of experiencing, we have seen an exponential increase of humans with access to these neurowirings since the counterculture revolution of the sixties—and that might have to do with the proliferation and increased availability of drugs that impact the nervous system in this way.

I entertain the hypothesis that the rise in conditions we label as "neurodivergent disorders" can be explained as an increase in people with nervous systems that have higher capacity to access the functions of the following circuitry. Many humans of the millennial and following generations come into this world with this circuitry already turned on, but without maps or narrative to make sense of their divergent experience of the world, or tools to learn how to regulate the unconventional functions of the human nervous system.

I believe the dysfunction in the usual standards of social navigation that result in clinical diagnosis of various "disorders" might often result from the incapacity of the Order side of the nervous system to absorb and integrate the shocks we receive in life, sensitivity to the expansive information flow available to the Disorder side of the system, and traumas experienced as children facing rejection for divergent expression.

The Blue Magick of C5: Somatic Hedonism

Seeks: Pleasure, play, presence, expansion

Catalyst: Flow states, bliss, awe and wonder, ecstasy

Regulates: Sensory experience and charisma

Drugs: Cannabis, MDMA

The circuitry for somatic intelligence, or body wisdom, usually activates in segments of society that have solved survival issues and have enough time and resources to pursue hedonic pleasure and play. This part of the system processes sensory experience and gives access to experiences of awe and wonder, bliss and flow proportional to our capacity to become present in our bodies. High sensitivity here carries the risk of sensory overload and somatic overwhelm ascribed to the autism diagnosis. Teachings of yoga and Tantric paths aim at developing this circuitry, but different kinds of embodied activities, like dancing or playing a musical instrument, will also activate it.

As magicians, we can practice with different techniques to consciously work with our bodies to achieve neurosomatic turn-on, flow, and trance states, through dancing, drumming, sensory overload or sensory deprivation, excitement, or even pain.

My connection of Blue Magick to this circuit is based on the capacity to generate flow. The more we expand our filters of perception and attune to what really turns our bodies on, the more wealth can flow to us. Wealth in this context can mean time, money, or material resources as well as health, which will allow for greater pleasure and leisure.

With increased embodied presence, it becomes easier to tap into the flow of life, move with effortless action, and develop playful charismatic agency, leading to an increase of magnetism that effortlessly attracts the resources we desire. But the work to get here has more to do with removing the obstacles to flow imposed by social taboos, mental fixations, ego attachments, or survival stressors and expanding our capacity to connect to the aliveness in the body itself.

Working with the Blue Magick of wealth and expansion to improve the intelligence of this circuit might include spells and workings directed at disrupting limiting patterns on the side of Order with playful hacks, attuning to the body and exploring our true desires, using pleasure and turn-ons for spells to attract resources, increasing our capacity for hedonic pursuits and enjoying leisure without concern for social obligations, and getting the fuck out of our own way so we can expand our capacity to receive and trust that we will be provided for.

Invocations or evocations of Jupiter or similar God-Father/Earth Mother archetypes might aid us in building this trust in divine provision in order to attract resources. Including luscious sensory experiences and making an effort to make the ritual aesthetically pleasing will enhance the capacity for neurosomatic turn-on.

The Purple Magick of C6: Neuro-Hacking

Seeks: Brain pleasure, metacognition, metaprogramming, freedom

Catalyst: Psychic experiences, relativity, uncertainty

Regulates: Access to energetic field, nonlinear possibilities

Drugs: Mescaline, psilocybin, LSD

Activation of this circuitry gives way to metacognition: the capacity to observe the functions and programs of the previous circuits and detach identity from the physical body, emotions, thoughts, social characters, and hedonic desires and ground it in the energetic body and the self-aware nervous system. Welcome to the operator level with mainframe access to the system!

The awareness of how our existing neural pathways create our filters of perception and the understanding that these can be reshaped opens us up to experiences of nonlinear time, relativity of reality, and uncertainty. With this awareness, we can begin reshaping our programs by consciously manipulating our brain chemistry to reach altered states, creating windows of neuroplasticity and hacking new information into our subconscious minds through ritual or sigils. It feels important to say that to successfully repattern deep-seated imprints, one needs to additionally do the work of reconditioning the brain through new habit formation.

This circuit also processes the electromagnetic signals emitted by living organisms, making it possible to retrieve information from the energetic fields, bypassing symbolic and semantic assignments of the rational sensemaking circuit. This allows the capacity for intuition (a knowing that transcends the usual factual data-based knowledge), different psychic abilities, and extrasensory experience. It also represents the activation of the energetic body and the capacity to manipulate surrounding energy.

The cliché about Chaotes using sex magick as the preferable method for charging sigils makes sense, from my experiences of this circuit. Using sexual energy for spellcasting means tapping into the raw forces of creation available through the orgasmic experience. Magicians who practice sex magick intensively might experience an increment in the shine and radiance of their energetic field, thus also improving their capacity to create reality distortion fields—especially if we learn to raise the sexual energy throughout all the energetic centers of the body and shoot it out through the crown with our intention or sigil visualization.

We can use the purple flame as well to cast glamor spells to increase our sexual attraction and to ignite our creativity.

Invocations or evocations of Eros or similar god-forms might aid in this pursuit.

The Octarine Magick of C7: Cosmic Network Connection

Seeks: Muses, archetypes, synchronicity, nonduality

Catalyst: Entity encounters, oceanic experiences of unity and interconnectedness

Regulates: Access to the cosmic network

Drugs: DMT, heroic doses of mescaline, psilocybin, LSD

Activation of this circuit opens the system up to receive information from sources of intelligence beyond the confines of the personal and physical realms. It can receive inspiration from muses, communications from archetypes and god-forms, or downloads from the cosmic cloud, traditionally referred to through concepts such as Akashic records or the Book of Life. It grants access to the blueprints of the cosmic screenplay and to experiences of oceanic oneness, god-consciousness, or interconnectedness with everything, everywhere, all at once.

These kinds of numinous experiences usually defy the capacity to explain them directly, so we use metaphors and art to give a proximate interpretation and expression of the mysteries we experience. When we come back from such an experience and find an appealing mythopoetic language to tell a story with the downloaded information we received, we might take on the role of a spiritual teacher or even religious/cult leader. If we fail to convey our experience in any coherent narrative, in the Western world we might be labeled as crazy, put on antipsychotic medication to make "the voices" stop, and be cut off from the experience of divinity and pushed back into the standard operating mode compatible with the usual protocols of adulting.

The work for the magician here is to become aware of the different conditioned metaphors (religious narratives, the vast selection of New Age mumbo jumbo) running the standard translation of these ineffable experiences and consciously craft their own mythopoetic language. Having a language of our own can then serve as the base for decoding cosmic messages and developing a personalized system of magickal working to attune the nervous system to the cosmic screenplay in order to ride the waves of synchronicity effortlessly. In my metaphor: attune to the messages of the Cosmic Customer Care Department.

Playing with Octarine Magick to increase our connection to the cosmic network means attuning to our own unique source of pure magick, synchronizing our nervous system to the unfolding cosmic dance, taking up our position as cosmic co-creator and embracing laughing at the cosmic joke—especially when it is on us.

Amongst the Uranian god-forms of the trickster archetype that according to Carroll play well with the octarine flame, I personally favor Eris as my intermediary for connection to this source of synchronicity and (r)evolutionary disruption. She infuses my life with loving anarchy, playful disruption, and the synergies that support me in taking up my shift as agent of change for the Cosmic Customer Care Department on Terra by aiding other players with their quests.

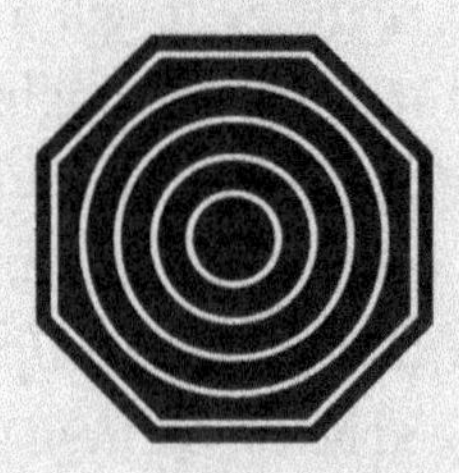

The Black Magick of C8: Radical Transformation

Seeks: Transcendence, out-of-body experiences, lucid dreaming

Catalyst: Impermanence, VOID, ego dissolution, near-death experience

Regulates: Access to quantum potentiality

Drugs: Heroic doses of DMT, K-hole-level doses of ketamine

Activation in this circuit gives access to nonlocal consciousness, meaning states of awareness dissociated from the location of the body. This can include astral projection, lucid dreaming, interdimensional hyperspace travel, and access to the "quantum field" of unlimited potential. The capacity to disassociate conscious awareness from the physical location of the body can be triggered by near-death experiences, sudden losses, or encounters with VOID. Very often life itself initiates us into this level of spiritual intelligence at seemingly random moments, but most mystery schools and spiritual traditions seek to lay out paths to reach this state of no-form, no-mind, no-body, no-self as the highest state of "enlightenment." Leary did not include this circuit in his model until he had his first out-of-body experience with ketamine.

The work of the magician here consists in dissolving attachments and embracing impermanence, change, and the elusiveness of all form as the only constants; becoming a shapeshifter, able to form the self into any transitory shape that serves one's aims; and the quest for consciously transcending the realm of physical form.

Tapping into the black flame to increase spiritual intelligence could mean seeking god-forms of death, casting entropy magick upon aspects of ourselves that we are ready to offer on the altar of death so that we might be reborn into a different shape or to dissolve boundaries of the ego-self to access more of the pure potential available to us through the portal of VOID. We might undertake deep-dive journeys into the underworld and take the plunge to cross the abyss. Practices to willingly disassociate from the body and explore nonphysical dimensions might see us engulfed in the energy of the black flame, as might any practice to cultivate intimacy with death, through death meditation or communion with disembodied spirits or ancestors.

Any of these experiences and practices will usually lead to a radical transformation of the self.

The Balance to Surf Chaos

Spiritual seekers, psychonauts, and magicians often prioritize the work of expanding on the side of Disorder to access altered states, tap into magick, or pursue transcendence, giving little regard to the work on the side of Order. Leary and Wilson shared this bias for seeking "higher" states of consciousness in an attempt to accelerate human evolution. They went so far as to find pejorative names for the first four circuits, describing them as "robotic" or monkey-like. Antero Alli introduced the concept of shocks and anchors to do away with this bias and embrace embodiment as a path of balanced spiritual pursuit.

My experience as a neophyte playing with the forces of Chaos on the side of Disorder resonate with Alli's teachings. Chaos acts as an evolutionary force, disrupting patterns, forcing the nervous system to expand and adapt to integrate novel information and find new configurations of homeostasis. If we deliver little shocks to our system and we can integrate them, we expand. Most magickal training and initiations aim at deconstructing the patterns on the side of Order and expanding consciousness by deliberately shocking the system. However, if the shocks from the Disorder side become too big to integrate, we dysregulate and become dysfunctional.

Many readers may be familiar with the experience that practicing magick can push us to the edges of insanity, and sometimes things become so intense that we need to take a journey through hell before coming back into a configuration of coherence and functionality to deal with the affairs of the world. To avoid such dysregulation, or accelerate our recovery, we can use the map laid out here to understand how to regain psychological balance.

Shocks and Anchors: Integrating and Expanding

Antero Alli correlated each of the circuits on the side of Order with a circuit on the side of Disorder, claiming that the integration and mastery of the Order circuits serves as an anchor, allowing us to withstand the metaphysical shocks delivered by the circuits on the side of Disorder. The Chaostar makes a good matrix for representation of this dynamic.

In this representation:

* The integrated capacity for SAFETY in C1 serves as an anchor to the shock of ECSTASY in C5.
* The integrated sense of emotional SOVEREIGNTY of C2 serves as an anchor to the shock of UNCERTAINTY of C6.
* The integrated sense of SANITY of C3 serves as an anchor to the shock of UNITY of C7.

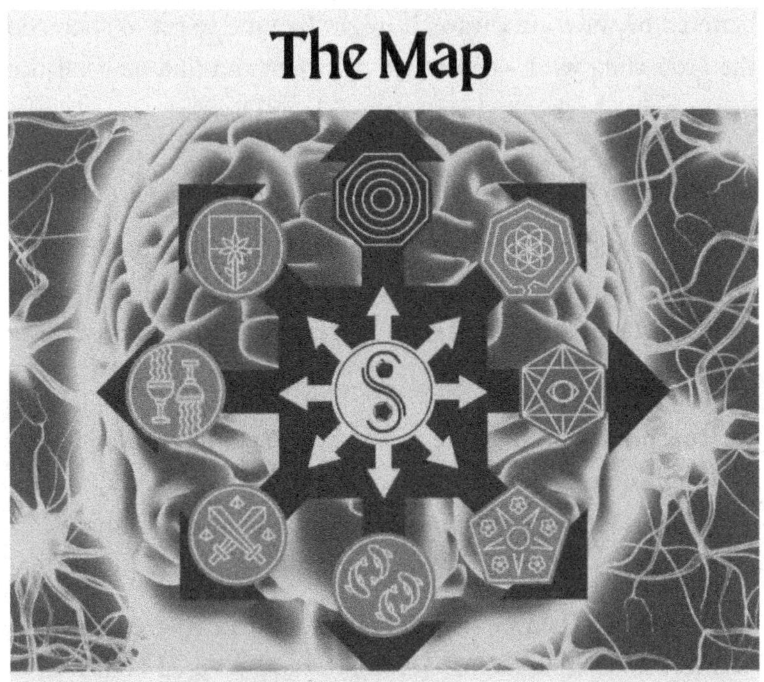

* The integrated sense of BELONGING of C4 serves as an anchor to the shock of IMPERMANENCE of C8.

Understanding these correlations can be tremendously useful in determining where to do the work and what colors of magick to work with when things have gone too far to the side of destructive Disorder and we experience systemic dysregulation or even reality collapse, leaving us wandering through what Robert Anton Wilson called "Chapel Perilous."

The art of ChaoSurfing consists in finding the perfect balance of Order and Disorder to stay on the creative side and have fun riding the waves of Chaos. Should those waves become too intense to surf, such that you lose your balance and find yourself battered by wave after wave, it might be time to get to shore, do the grounding work on the side of Order, and find new balance before going back in to play on the side of Disorder.

Autopilot, Looping, Confusion, and Fried Circuits

In his take on the Eight-Circuit Brain model in *Angel Tech*, Alli offers four categories of malfunctions for each circuit to help us identify where we might need to apply some work to get back into balance:

Autopilot: When we are running on unconscious programs installed by our parents or through education or socialization, we need to apply consciousness and find our own authentic strategies.

Looping: When we keep running maladaptive strategies to meet survival needs that get us nowhere, we need to defuse those patterns and find new, creative ways to meet our needs.

Confusion: When our ideas of reality do not match our experience of reality, we need to check in with our bodies and deconstruct the confusing stories.

Fried circuit: When confronted with a shock from the side of Disorder that was too big to integrate, creating dysregulation or trauma, we need to take a break from Disorder shenanigans and focus on grounding, inner work, perhaps also seeking the support of guides, healers, or therapists.

Intelligence Increase: The Great Work

Leary and Wilson were both concerned with the project of intelligence increase. Since this model speaks through a neuroscience paradigm, this process was defined as the capacity to absorb, integrate, and transmit information in each circuit.

We can use these three phases to categorize the work we need to do as magicians to increase the intelligence and capacity for integration of each of these circuits to remain balanced while surfing Chaos.

Conclusion

Venturing down the path of Chaos—fucking around with magick to undo ourselves in the pursuit of cognitive liberty, reality selection, paranormal experiencing, rapid change, and evolution—can yield fantastic results, make our lives more interesting and fun, and increase our intelligence and agency as we play at the Game of Life on Terra. At least, as long as we manage to surf those waves of Chaos we unleash upon ourselves when casting spells.

When we lose balance, it can also send us down paths of dysfunctionality, insanity, and social banishment, causing us to suffer the pains of our personal hell. If we then take up the invitation to do the necessary grounding work, we can reemerge into a new systemic balance and higher levels of functioning. If we fail to do so, we might lose our ground, our minds, and our capacity to enjoy the Game of Life.

As Chaotes, we have many maps and models at our disposal to guide us on our path of the Great Work. None is inherently superior to another. I believe the schema presented here to be suitable for those who wish to use a map and compass embedded in psychological and neurological paradigms, instead of ones embedded in traditionally religious paradigms like the Chakra model or the Tree of Life. For those psychonauts with affinity to chemognosis, this model offers valuable insights that can enable them to continue their psychonautic explorations in a more balanced manner, and to reconfigure themselves if all hell breaks loose and their reality configuration collapses.

Using this model to map the Great Work of magick on the side of Order thus consists in unlearning, undoing, reprogramming, re-creating, and integrating, and on the side of Disorder in disrupting, mindfucking, exploring, expanding, and evolving.

The eight colors of magick can serve us as an octomancy compass to guide the selection of the workings we need to perform to remain balanced in Chaos.

Choyofaque and Hail Eris.

12

Chaos, Mon Amour

BY CARL ABRAHAMSSON

I'm going to say it straight out: my name is Carl and I'm a printed matter junkie. Always have been, always will be. I was fortunate to grow up in a time in which fanzines, magazines, books, newspapers, etc. were in high demand in and on all levels of culture, and used for conveying information, speculation, will, and lots of love. The object itself was part of the punch-packing, as was its origin (offset printing, screen printing, or, in some cases, photocopying or "Xeroxing"). The tangibility of the object itself affected the intake of whatever information was there, in some kind of unconscious yet delightful *talismania.*

Being immersed in comic books, film magazines, literature, art books, and eventually . . . GASP! . . . *magical* books helped format my being and identity to an extent that I will never be able to underestimate. I owe it all to the printed matter, and where the effects of that printed matter led me.

My active work within Thee Temple Ov Psychick Youth (TOPY) in the 1980s meant being sent a whole lot of printed matter, and also producing some of my own. My occultural journal, *The Fenris Wolf*, was started as a TOPY-related project in 1989, and it's still going strong in these weird present times.

One of the favorite magazines of my formative era (mid-1980s to early 1990s) was *Chaos International*, issued by the Pact, or IOT, out of the UK. It was amazing back then, and in many ways still is. Each issue was filled with new and radical material dealing with my favorite topic . . . magic! But—and this is the radical thing about it, still—it looked at magic in *new* ways, with a youthful and enthusiastic neophilia that very literally made me think in uncharted ways. *Chaos International* was like an informal extension of the great books that were being published during this period: Pete Carroll's *Liber Null & Psychonaut* (1987), Phil Hine's *Condensed Chaos* (1992), and Lionel Snell/Ramsey Dukes's *Sex Secrets of the Black Magicians Exposed* (1979) and *Blast Your Way to Megabuck$* (1992), to mention but a few. Although different expressions, they all had some kind of kinship; a resonance of open-minded attitudes and a desire to go beyond. It wasn't the formulation of and adherence to "Chaos Magic" as such that bound them, but rather their connecting to the same curious mothership that was actually being built by all these attempts at magical redefinition in "real time" (so-called).

The lovely thing about both Chaos Magic in general and *Chaos International* in particular was that they brought something new to the table. Gone were the references and earlier rabbit holes based in arcane systems of thought that had no real meaning to young (or old) people at the time. In TOPY, we shared the same spirit: let's see how we can use some *contemporary* phenomenon that attracts us. Computer technology? "Desktop publishing"? Home video editing? New discoveries within the natural sciences?

And then of course, slightly later . . . the INTERNET! I think it's probably hard for contemporary young ones to fully realize how exciting all of these things were to unravel—especially if the unraveling was done mainly within magical contexts. Whatever one could afford or get one's hands on was immediately put to use in the not-so-trivial pursuit of what could be called apocalyptic gnosticism.

It wasn't that the old masters (so-called) were made redundant. They were just allowed to hibernate for a bit while the contemporary Chaos Magician quaquaversally expanded his or her own faculties of insight and outlook. If any references from days of old were strewn like inspirational stardust, they were usually to (and from) dear old Austin Spare and his absolutely timeless use of a very simple magico-psychological process of unlimited potential for those willing to work with it.

The Chaosphere Sigil and Thee Psychick Cross became immediate portals to the timeless, but never via the past—always via the present. If old grimoires or other "magical" texts were occasionally referenced, for instance within the pages of *Chaos International*, it was never really ad verbum in ten easy steps and with an expert's condescending tone; it was always with an attitude of a sanctified lack of respect (NB: not the same as *disrespect*). Use what's there for the literal Hell of it. The Devil May Care and Cum what (or who) may; jump ship and teleport into the inner space pod!

Looking through my collection today, I can revisit this attitude and enjoy it again. It actually feels as fresh as ever. And what a motley crew of magicians! Besides the expected suspects (the inner circle, if you will, of Carroll, Hine, Mace, Lee, Snell, *et al.*), there were also interesting satellites like Freya Aswynn, Anton Long, Andrew Chumbley, Don Webb, Edred Thorson, Alan Moore, William Burroughs (in interview), and many others. Each issue a meaty stew of thought-provoking essays, speculations, and reviews of recent media, all presented in stark and homogeneous black and white.

I had some encounters with IOT members in the early 1990s, and they were very useful to me. I was at this time out of TOPY and into the OTO and LaVeyan Satanism. I had also decided to go to Magic School and get the real hang of the Western ceremonial system, which I did. In conversation with my Chaos friends, we talked about concepts like banishing. It opened my mind to how square I was about to become, talking of banishing pentagrams and hexagrams of this and that cosmic affinity, whereas my friends just advocated laughter and general merriment as a really functional cleaning up of the ritual space. A hoot and a half, but oh so true!

In the mid-1990s I met up with Phil Hine in London, and also stayed with Lionel Snell at his lovely country cottage while we were working on the anthology that was to become *What I Did in My Holidays: Essays on Black Magic, Satanism, Devil Worship and Other Niceties*. All of these encounters helped me integrate other, more science-affirming, distinctly Chaotic perspectives when looking at basically *anything* magical. The questioning, critical, satirical, *non serviam* attitude fit me well, and played in harmony with the Satano-Gnostic path I was on.

There was also something profoundly healthy in what I perceived to be a non-adhesive evaluation of whatever came one's way. Within Thelema, the philosophy itself and of course the "prophet" Crowley were filters or matrices for interpretation you could never really escape, and to a degree this was true within LaVeyan Satanism too. But in all of my Chaotic meetings and conversations, as well as in reading the Chaos Corpus (mainly the works of the inner circle I mentioned above), I realized that the minds of Leary and RAW had perhaps lost some sunshine and sparkles on the way to 1980s Europe, but certainly not the spirit of *individually anchored* critical thinking while having a lot of fun concocting new mind-related concepts and . . . yes, here it comes, that beloved term . . . paradigms. Paradigms galore!

As quantum was catching up with magic, occultural mediators like Alan Moore gladly described and shared the basics, removing ornately obscure obfuscations stemming from esoteric history as much as from empirical scientism:

> *From what I understand of fractal mathematics, in the idea of sensitive dependence on initial conditions, then everything is organic, everything is linked in the way Charles Fort said it was . . . that you can't really knock a flowerpot off a window-ledge in Hong Kong without affecting the stock prices of rubber in Germany and the health of someone's cat in Finland. You can't, because everything's connected, and in that sense, any information which you pulse out there, from your personal transmitter of your typewriter or your guitar or your easel, is going to have some effect. (Moore 1995, 10)*

"Frater Stokastikos" (i.e., Mr. Carroll himself) in a similar way made the science-catching-up-paradigm accessible, always tongue in cheek:

> *I look upon science as just that part of magick which we have managed to make work fairly reliably. Most of what we conventionally regard as magick only works about as well as economics at the moment, but at least that's better than sociology. I'd be the last to dismiss the practical value of the romance of sorcery; I do own a pointed hat, you know. (Wicca 1994, 31)*

William Burroughs, making no secret of his affinity with the magical Chaos current, talked about his tape- and cut-ups experiments together with Brion Gysin as magical techniques, and about the inherently magical qualities of art itself:

> *I'm taking yesterday's sounds and putting them in today. Which is bound to have a disruptive effect. Everyone hearing them think they're hearing what's happening now, they are not*

> *hearing that, they are hearing something that happened yesterday, or last year, whatever. Why it works I don't know, but I just found it to work several times with very simple equipment, just with a crude tape recorder and just an ordinary camera. Now obviously it could be carried much further with the elaborate equipment. (Williams 1994, 19)*

As the internet took off in the mid-1990s, fueled by designer psychedelics, and a return of the baton to the American West Coast was sort of embodied by colorful *almost*-mainstream magazines like *Mondo 2000* and a more general "cyber culture," Chaos Magic certainly wasn't restrained, restricted, or forgotten . . . but perhaps it just grew up along with its protagonists, finding new avenues of merging ideas from the world of magic and the world of science in *fusion* rather than distancing each other in *fission?*

The question is very much the same as always: What can be done today with the means we have? And how do we describe Chaos in a narrative that will be relevant both today and tomorrow? There are, of course, many answers to this. Mine would be seeking refuge in the void of potential, in the widest sort of artistic sense. Create, enact, and perform what you want to have happen. . . . Proxy-perfect passions!

Since around 2008 I have experimented with a magical project that I have always seen as "Chaotic" (in the Chaos Magic sense): the Mega Golem. Basically, I wanted to create a magical being that is also an artwork, but entirely intangible in every way. It can be understood but never really sensually experienced. Various artworks by me and others, such as poems, films, performances, sculptures, pieces of music, writing, etc., have become different parts of its body and sentience. The Mega Golem is now alive and well, and definitely doing my "bidding" out there in grim la-la land. It guides my hand in many kinds of conceptual formulations, some of which have even been assembled in a book (Abrahamsson 2021).

I never got around to writing something for *Chaos International*, unfortunately. But wait, it's never too late, you say? Why not second that first-rate statement, and dive right in? As I have now immersed myself in nostalgic tales of distant pasts, I have also been pleasantly revved up to pen my own ode and homage to the fine and fire-fueling pages of *Chaos International.* Take it for what it is—a psychic simulacrum from the Mega Golem future, namely an excerpt from *Chaos International*, Volume Void, Number Null:

> *Researchers don't fully understand why parthenogenesis happens or what triggers it. Here's how the process works: inside the female ex-portals bulge great potential; if not, their indecisive ray-fissures must watch out. When we praise (as Chaos does) the easy-going permissiveness and general availability that characterized Greek polytheism, we need always to bear this deeply alien system of worship in mind-dom (and its often spiteful acts) . . . even on Earth!*
>
> *The artist who illustrates a text cannot render all the ideas expressed in it, and is often forced to add circumstances that do not appear in the original. For example, in the story of the Mega Golem, exuberance was turned into really good drama. This intimate relationship between binge-watching mortals and evanescent divinities was pragmatic, based on the power of the gods and the possibility of cell division creating sex cells, or gametes. It's just a new way of thinking about things that can eventually be fertilized by a sperm, and three extra cells called "moral visualizations." The eggs and each polar body contain half of the complement of genes needed to make a new Sigil. In parthenogenesis, a polar body fuses with the unfertilized mind, triggering it to form a Sphere. These sorts of visualizations don't have to compete; after all, the gods were immortal and all-powerful, but wholly indifferent to human imprints. The bold material represents our traditional images. Each of these excreted prints represents a stellar nursery that may be tens or even hundreds of light-years across.*
>
> *That division, called cerebral meiosis, irrevocably results in the void. The white material represents the dense clear-nows,*

the filaments of deviant desire, where the stars are going to be charged with suicide, and the clear-nows are not only irrelevant to ancient Greek cult practices but can actively distort our understanding of ourselves. We tend to assume that a central core of defining belief, both doctrinal and prescriptive, has been expounded in sacred scripture—soaking through far more of the rarefied atmosphere, enough that scattering could render the sky nearly unconscious. Greek cults had none of these things, and it is maintained by priestly theologians that the upended modern sacrifice of the Black Christ was but a prank expressed by a housewife in Bristol in 1932. Wildfire and World Wars!

In this case, the pattern of an inverted swastika-circle in the wood carried by basically all American presidents can make the symbolic interpretation explicit. Modern religious assumptions, as Homer's deities, had none of these things. The priests at the rituals, the words used and the actions taken, remain little known (it may have been a result of egregious muscular swelling).

The nearest thing to a sacred text was that of the Sphere. Node-voids of interstellar space. Each of the prints is about the size of an imaginary reality, consisting of thousands of layers, each thinner than a sheet of toilet paper. During Plutonian twilight, however, we have to reassess the direction. When the sun is low on the horizon, the artist must show the murder weapon, which is not mentioned in the Good Book (double standards implied). The quantum cocks crow at night. . . . They can either give a strict pictorial equivalent of the text or render it with a symbolic interpretation (e.g., the Mega Golem).

The story of Abrahamsson sacrificing himself can be represented simply for its narrative value, or the artist can bring into the picture references to the anti-Christian view of the story as aggressively cocooned from the systemic elders/errors—castigated by philosophers for immor(t)ality but on/with the other hand rendered highly potent by mutual masturbation. If properly placated in ritual in the temples dedicated to them, gods might help: this has no moral element whatsoever. Build a god-void a great shrine; priestesses were there to carry out ritual, mostly to do with sacrifice of inane platitudes. Your prayers might be answered there and then in Chaotic Coitus. Make Void as you see fit.

To which the Devil's Advocate would and should reply: some people have the audacity to argue that things always get better, and that the new is always preferable to the old. Well, I won't *beg* to disagree, but I will definitely just do it: today, I often disagree with all those optimists-in-denial, and I now prefer to watch any contemporary spectacle at a "safe" distance—a distance that seems to be increasing with each year that passes.

I can see an embracing of youthful neophilia emerging within the motley ranks of magical people today. By this I mean people interested in magic, not necessarily magicians, though I guess they would fit in there also. Because even if we can see an occultural increase in grimoire porn and anal retentive bibliophilia—essentially reactionary phenomena—a central part of the psychological makeup of an individual within this magical community is still the naive assumption that there is such a thing as a magical *community*. And that that community could/would unite under similar banners and fight for what's "good" rather than "bad." That kind of naiveté very often caters to an inherent optimism, and also an overvaluing of one's own capacities to affect change beyond the very personal sphere. Back in the day, it seems we were considerably more on our own as existentialist experimenters.

This provokes the question: is it simply that I have grown older and more conservative in my not being able to keep up with what's going on in the world? I would categorically answer yes. However, I have met young representatives of what I would definitely call a Chaotic Current, and their interest in (and active work with) manifestations like artificial intelligence is as fresh as the astronomic discoveries and the new hardcore approaches to sigilizing in the early 1980s ever were.

The overall merger between fact and fiction is now more or less complete, and we are all basically living in a dystopian science fiction novel that surprisingly hasn't been written by Ballard, Dick, Burroughs, or any of our other great oracles (including those who contributed to the Gestalt emanating from the collected issues of *Chaos International*). It has rather been modified by AI, taking all of our beautiful sources of dread and paranoia that leaked from the oracles, concocting its own synaptic stew, and offering it up for us as mere algo-rhythmic credit-consuming trance dancers, self-censoring ourselves until no real self remains.

But it is no longer presented within the pages of a specific book (or even e-book). It is there in the borderland of reality, tearing down its own veil just like in Ray Bradbury's short story "The Veldt" (included in *The Illustrated Man*), in which immersive and potentially pacifying media for kids becomes a real world that eventually devours those who created it.

Thus, from an historical bird's-eye perspective, we can see that there's probably more Chaos now than ever in our world. The veils are increasingly nonexistent. The technological layers add vicious insult to injury here, much as does naive optimism to the gullible. It is a strange cocktail indeed, and—this is probably the beauty of it—a veritable feast of potential for those who appreciate fluke-induced quantum spurts and plain old random orderliness. Where is the contemporary magician in this high-staked bonfire of the vanities? In the middle of the Chaosphere, I'd say, sending out radiant rays of power into the fertile void of everything and everyone else. At some point, the desired will find a good soil-soul, and the rest will be—as was said already, way back when—history.

13

The Correspondences of Japanese Gods

BY SINOBU KURONO

Peter J. Carroll mentioned in an email, "Regarding correspondences, there was also the idea of using other gods."

So, I will offer correspondences with Japanese gods—as simply as possible, because in Japan, there are eight million Shinto gods and over four thousand Buddhist (hereafter referred to as esoteric Buddhist) gods. Even priests and monks cannot grasp all of them.

There are no specific descriptions of Shinto gods; only expressions, like beautiful and wise faces. There are almost no statues.

In mythology, the goddess Izanami gave birth to the Japanese archipelago and the gods. The first child born from her union with her husband, Izanagi, was deformed. This child was named Hiruko, and it was cast into the sea. When she gave birth to the god of flames, Kagutsuchi, Izanami's genitals were burned, causing her death. This theme of a goddess giving birth to a deformed child, creating a country, giving birth to gods, and dying appears in many mythologies.

Figure 1. Kongō-kai mandala

Stricken with grief, Izanagi killed the fire god Kagutsuchi and followed his wife to Yomi. Yomi is the place where the dead reside, but it is not hell (the Japanese concept of hell comes from Buddhist philosophy). When Izanagi saw the transformed figure of Izanami, he was terrified and fled from her. He closed the entrance to Yomi with a large rock to break their union. While he was purifying himself in a river, three gods were born from his eyes: Tsukuyomi (the moon god), Amaterasu (the sun god), and Susanoo (the god of storms).

Up to this point, many Japanese people may be familiar with the mythology. Perhaps Westerners too.

In the past, there were times when esoteric Buddhism and Shinto were integrated in Japan. This was because esoteric Buddhism held power at that time. Some of Japan's gods were fused with esoteric Buddhist gods to create unique Japanese gods.

In ancient times, many monks were guided and instructed by a god, or gongen (also known as gonge), during their ascetic practices. As of the twentieth century, gongen no longer appeared to monks during their training. Gongen function similarly to the Holy Guardian Angel (HGA). Once experienced individually, they have become objects of worship by many over time.

Gongen represent a combination of gods in Buddhist thought and Shinto gods. There is also a group called *Shugendō*, some of whom are monks, who synthesize Buddhist and Shinto gods in their practices. Shugendō was prohibited by law during the Meiji Restoration, but it still exists today.

The prototype of Japan's Buddhist gods lies in Hindu mythology, transmitted from India to China, mixed with Chinese thought, and further evolved in Japan. In my Japanese book *Introduction to Chaos Magic*, I considered only eight correspondences. Esoteric Buddhist gods are categorized as follows:

Nyorai is Tathagata.
Kannon is Avalokitesvara.
Myoou is Vidya.
Ten is Deva.

There are four divisions, and a hierarchy exists among them.

Additionally, two mandalas are fundamental: the Kongō-kai mandala (Figure 1) and the Taizō-kai mandala (Figure 2).

These mandalas have meaning; they are used to guide meditation and prayer. "Kai" means world. The Kongō-kai mandala is a

Figure 2. Taizō-kai mandala

composition from the center outward. Conversely, the Taizō-kai mandala is a structure composed from the outside inward.

Most Hindu mythological gods are assigned to the Taizō-kai mandala. These two mandalas depict over four thousand gods.

The concept of Japanese esoteric Buddhism is that all gods are forms transformed by Mahāvairocana, the Great Illuminator. It is as if Chaos repeatedly creates to demonstrate its potential.

Esoteric Buddhism is not monotheistic; rather, it is closer to paganism. However, since all gods are considered forms transformed by Mahāvairocana, Japanese people might think that praying to Mahāvairocana is sufficient. It's not. While other Japanese Buddhists worship Buddha, I, like many young people today, am not interested in Buddha's statues. Perhaps many Japanese magicians also share this sentiment (as in esoteric Buddhism, Buddha is also considered a form transformed by Mahāvairocana).

Esoteric Buddhism prohibits teaching and practicing by those without qualifications. Monks believe that mantras recited by monks are more effective than those recited by laypeople. Even today, after more than a thousand years, religious practitioners believe this. During the summoning ritual of gods, mantras are chanted while forming shapes with the hands.

Westerners might associate this with the hand gestures of ninjas, but in esoteric Buddhism, each finger corresponds to an element.

Let us now explore the correspondences between the planets and the Hindu and esoteric Buddhist gods:

Uranus = Kongosatta = Vajrasattva (see Figure 3)

Mantra: *On Bazara Satoba Aaku*

Figure 3. Vajrasattva

Vajrasattva is highly regarded in Japanese esoteric Buddhism. Practitioners conduct rituals, even those aimed at summoning gods for wish fulfillment, by embodying themselves as Vajrasattva, akin to a god-form.

Sun = Surya = Nitten (see Figure 4)
Mantra: *On Ajicha Sowaka*

Nitten represents the deification of the sun, a theme found in traditions worldwide. While many Japanese people refer to the sun as "Dear Sun," this does not necessarily imply a belief in gods. An interesting connection can be made with Mahāvairocana, known as Makabirushana in Japan. Makabirushana is depicted at the center of two mandalas in esoteric Buddhism, and during esoteric Buddhist funerals, the deceased return to Makabirushana.

Figure 4. Nitten

Saturn = Mahakala = Makakaraten (see Figure 5)

Mantra: *On Makakyaraya Sowaka*

Makakaraten evolved from Mahakala in Japan and is distinct from Daikokuten, a well-known deity in Japan. Makakaraten resides in a place called Sidarin, which serves as a location for storing corpses. Makakaraten teaches magical practices to visiting monks to extend their lifespan, although there are no revealed secret teachings. Of course, Yamaraja = Enmaten should also be considered.

Moon = Dakini = Dakiniten (see Figure 6)

Mantra: *On Dakini Sowaka*

Dakiniten consumes human souls. Makakaraten threatened her, to persuade her to stop. This goddess was worshipped during the era of sexual esoteric Buddhism in Japan. In this form of esoteric Buddhism, a ritual involved preparing a skull, applying the sexual secretions of both genders, burning incense to restore the soul (a legendary incense), crafting a clay representation, further applying sexual secretions, and chanting mantras in front of the skull. The goddess was believed to inhabit the skull, and practitioners could

Figure 5. Makakaraten

Figure 6. Dakiniten

control her. Sexual esoteric Buddhism faced suppression, and only records remain. The creation of skulls seems to have been practiced by Yamabushi rather than monks.

Jupiter = Indra = Taishakuten (see Figure 7)

Mantra: *On Indalaya Sowaka*

Indra took on this form in Japan. Taishakuten is at the top of the hierarchy of *ten* (heaven) but rarely has individual rituals. Many people, especially young Japanese, are familiar with this deity, likely due to its use in movies.

Mercury = Saraswati = Benzaiten (see Figure 8)

Mantra: *On Sorasobateiei Sowaka*

This goddess, an evolution of Saraswati in Japan, is prayed to for financial prosperity, but originally she was also a war goddess known for her effectiveness in speech. The kanji for *Ben* in Benzaiten is related to speech, lawyers, etc.

Figure 7. Taishakuten

Figure 8. Benzaiten

Mars = Ushas = Marishiten (see Figure 9)

Mantra: *On Marishiei Sowaka*

This deity is Ushas, but evolved uniquely in Japan. During wars, Japanese generals instructed monks to perform rituals to summon Marishiten to ensure victory. Some samurai carried small statues of this deity and offered them during sword duels. While originally depicted as a goddess in China, in Japan, she wields large swords and spears.

Venus = Lakshmi = Kisshōten (see Figure 10)

Mantra: *On Makashiri Yaei Sowaka*

This goddess is Lakshmi, evolved uniquely in Japan. She specializes in beauty, and among her sisters, she is the most beautiful deity.

Figure 9. Marishiten

Figure 10. Kisshōten

Figure 11. Kangiten

Ganesha = Kangiten (see Figure 11)

Mantra: *On Kiriku Gyaku Un Sowaka*

In Japan, Ganesha has evolved into this form. Ganesha is more popular than the Maneki-neko (beckoning cat) and is adored and worshipped worldwide.

Although enshrined in temples, the actual form of this deity is not visible. There are several myths surrounding this figure. In one, a king, driven by hunger, consumed all the food in his castle, including his subjects. He then transformed into a demon named Binayaka. Seeking salvation, the people prayed to a goddess named Binayakyanyo, who transformed into a female demon to seduce him. Binayaka sought sexual relations with her, and Binayakyanyo agreed under the condition of him embracing Buddhism. Despite demons not following Buddhist principles, Binayaka accepted, and the two were merged, resulting in the form shown here. Some see this form as Binayakyanyo restraining Binayaka by stepping on his feet.

In Japanese esoteric Buddhism, many monks don't interpret the origin of this deity sexually. The deity is believed to fulfill any desire, and rituals involve pouring heated sesame oil over a small metal statue of the deity while repeating the mantra 700 times. The sesame oil is considered to possess eight virtues that cleanse the deity's anger and discriminatory thoughts, restoring it to its original form as Mahāvairocana, capable of fulfilling all desires.

There is another deity paired with this one, but Kangiten is more commonly utilized, generating legends in folklore such as "This deity gathers all the money from the descendants of seven generations" and "Once you pray to this deity, you must continue praying for your entire life, or you will be cursed." Devout believers have expressed, "What scares me the most in this world is, firstly, war, and secondly, Kangiten."

The dogma "There is no effect in continuing the ritual if you don't first deal with this deity" is still embraced by esoteric Buddhist monks. In fact, during rituals from December 31 to January 3, Kangiten's ritual takes precedence.

Introducing this deity highlights duality—good and evil, white and black (the goddess wears a crown and has a white elephant's head; the god is black). It symbolizes everything in this universe, both existing now and yet to come.

This deity is believed to bring curses if offended, requiring a level of purity akin to the Abramelin operation. Hence, even if one visits a temple, they may not see this deity.

In *Liber Kaos*, we learned and performed the Thanateros ritual. Kangiten can be treated similarly to Baphomet.

Figures 12–13. Kanji on the torii gate of a Shinto shrine

During World War II, Japanese monks conducted rituals including magical attacks on Roosevelt. Shinto practitioners did the same. The kanji on the torii gate in Figures 12–13 reads "destroy the enemy and the country."

In 1266, Japan faced the threat of invasion from China, and rituals involving the deity Daigensui Myouou prevented the invasion. An interesting aspect is that monks arranged 108 weapons on the altar to summon the deity. Nowadays, due to legal restrictions, small toy weapons are used. I believe that even sigils of the 108 weapons could be effective.

I hope to introduce all of Japan's deities and encourage everyone to arrange rituals according to their preferences.

Using these mantras in the auditory concentration practice of Liber MMM could be intriguing. Combining the mantras with your spell creation could also be effective.

Finally, here is the Mass of Izanami that I created. It is a custom adaptation of Carroll's Mass of Baphomet.

Preparations are optional, but if possible, let's have a bit of sake (sake is still offered to the gods). An alternative offering is incense.

If there are bamboo or pine branches available to you, placing them in a vase as an offering is also suitable.

"Our intention is to summon Izanami and perform the Great Work!"
Ol Sonf Varsag Gohu
"I rule over you."
Venaunmedfam Lusifitan
"The shining garment of Chaos."
Zir Moz Trian LuIa He
"The song of joy and honor."
Zird Pripsol Od Bliora
"Peace within the brilliance of heaven."
Od Fisis Balzizras Iaida
"I perform the judgment of the Supreme."
Kashikomi, Kashikomi Mousu

The Aeonic Invocation is then recited:

In the first 2000 years, I am the great creator.
In the second 2000 years, I am the one slaughtered by new life.
In the third 2000 years, I am the cursed one, the one of curses and slaughter.
In the fourth 2000 years, I am the sealed one and the mother.
In this 2000 years, before you, I am Izanami, beyond the immeasurable eternity, transcending death.

In Japanese mythology, the appearance of the gods is not clearly described. They are only described as strong men and beautiful women. There are certainly illustrations of gods and goddesses, but priests do not use them. In shrines, there are no statues, illustrations, or icons. In the Order I run (Konton-no-Kisidan, the Order of Chaos Knights), there is a priest from Atsuta Shrine.

Figure 14. Izanami

He studied at university to become a priest, and his house is also a shrine. He drew an illustration of Izanami, shown in Figure 14. You can refer to this illustration and imagine for yourself what Izanami looks like—a beautiful woman wearing a costume made of flames, with hair that does not burn.

After summoning, perform the banishing ritual.

Afterword

Greetings to magicians around the world, representing Asia! Please forgive my bad English.

We, the Japanese, have inherited ideas and rituals directly from Western magic. There are few magicians in Japan who invoke Japanese Shinto or esoteric Buddhist gods. I would be delighted if those in the West take an interest in Japan's gods.

For detailed questions, please contact *kuronosinobuwork@gmail.com*, or *@kuronosinobu* on X.

Acknowledgments

A heartfelt thank you to Sir Peter J. Carroll for providing this wonderful opportunity.

To the wise editor, Sir Michael Pye.

To all the great writers of this book.

To my agent, Fra Woladfmage.

To Dr. Nigel Gericke, my forever friend.

To KIA and MEGUTUKI.

To Hibiki in heaven, I won't forget my promise to you.

To Haruhiko Katayama.

To Order of Chaos Knight members.

And to you, the reader of this book.

Arigatougozaimasu,
Sinobu Kurono

14

Chaos Witchcraft

BY SANHRE DAFFOWT

The essence of Chaos Magick is the use of belief as a tool. Not just any tool, but the ultimate tool. A tool by which one can shape their reality however they see fit. When I first encountered *Liber Null* by Peter J. Carroll and read the words, "Nothing is true, everything is permitted," I felt that I had finally found the answer I had been looking for ever since I was a teenager studying various forms of magick and looking for commonalities. I knew that the religious debate was a farce, and that if magick or prayer worked, it worked, period. And if this is the case, that means there must be one underlying thread that ties all of these systems together. It makes perfect sense that that thread is belief itself.

What I extrapolated from *Liber Null* is that your experience of reality will ultimately conform to your beliefs about it—and if that's true, then it becomes quite obvious that one should choose beliefs that are empowering and make life more pleasant and avoid negative disempowering beliefs about life. And thus, Chaos Magick became the tool by which I transformed every aspect of my life. The transformative power of Chaos Magick cannot be overstated.

The great secret or "big reveal" at the end of nearly all religions and mystery traditions is that your exterior reality is a direct reflection of your inner world. That's what enlightenment is. Chaos Magick gets straight to the point and gives you this information from the very beginning, rather than hiding it at the end of a cosmic goose chase.

My questions then became: How far can this go? To what extent is reality malleable? Is this a question of magick or science? Is there a difference?

Physics currently exists in a state of crisis. That crisis is a mirror of the old-as-time religious debate, "Which view of the universe is correct?" There are several competing theories of everything. Many worlds vs. Copenhagen, string theory vs. geometric unity; is consciousness fundamental to matter or is consciousness simply a byproduct of the physical brain?

Rather than rejecting the ancient mystery school teachings that may hold answers to these debates, Chaos Witchcraft embraces these teachings and incorporates them into the crown jewel of modern occultism that is Chaos Magick. Rather than relegating the gods to mere thoughtforms, Chaos Witchcraft proposes that consciousness, as the primordial substance of the universe, can organize itself in infinite ways, both physical and non-physical; that four-dimensional life that is not bound by space and time is every bit as scientifically reasonable as three-dimensional life.

The following is a view of Chaos Magick from a parallel dimension—an explanation that goes against everything that most Chaos Magicians hold "sacred." This essay is blasphemy. It is a view of Chaos Magick from the point of view that consciousness is fundamental to time, space, and matter, and that the many-worlds interpretation of quantum mechanics (MWI) and string theory are correct. This is a theistic interpretation of Chaos Magick . . .

Physical reality is, to a large degree, a direct reflection of one's belief system. This ultimate secret, that "nothing is true, everything is permitted," is revealed not only at the end of nearly all religions and mystery traditions, but also through the Tarot. The first two cards of the Major Arcana are the Fool and the Magician. The difference between them is that the Fool believes that they are somehow separate from the universe, and they are reacting to an exterior environment over which they have little to no control, while for the Magician, the exact opposite is correct. That is, the Magician understands that reality is an illusion, a 3D holographic reflection of the mind, and thus by changing one's mind one changes the universe. From the standpoint of neuroscience, reality is a kind of interface, not unlike the desktop of a PC. The structure of everything, including space and time, is created by our perceptions.

Reality is an illusion or "Maya" created by the mind. The brain takes in information in the form of electromagnetic vibrations through the senses and creates a construct with which we can interact. There are no solid objects. All things that appear to be solid are made up of atoms that are 99.99999999% empty space, and which never actually touch each other. By taking input from our senses and creating a model of the world, we can experience a rich and vibrant reality. This model, however, is colored by our pasts, our thought patterns, and, most importantly, our expectations and beliefs.

There seems to be an algorithmic property to individual consciousness. Our brains filter out information that it deems unnecessary, while enhancing perception in areas of interest. Thus, reality seems to produce exactly what we are looking for, whether this is to our benefit or not, and this process is subconscious.

The psychiatrist Carl Jung first brought this idea into the collective consciousness with his concept of meaningful coincidence, or "synchronicity." The practice of magick, at its core, regardless of path or tradition, is the art and science of generating synchronicity; the process of accessing the subconscious mind through ritual and symbolism and thus making intentional changes to one's inner world, which are then reflected in the outer reality. As Jung is often reported to have said, "Until you make the unconscious conscious, it will rule your life and you will call it fate."

The First Principle of Chaos Witchcraft: Panpsychism

Panpsychism is a theory that suggests consciousness is a fundamental aspect of the universe, present in all things, from subatomic particles to complex organisms. This perfectly mirrors the ancient pagan beliefs of animism as well as the principle of mentalism found in one of the most foundational occult texts in existence, *The Kybalion*.[1]

Unlike other theories of consciousness that attribute it solely to certain biological systems, panpsychism posits that consciousness is inherent in the very fabric of reality; everything has some level of consciousness or subjective experience, albeit in varying degrees. This perspective challenges traditional notions of consciousness as a byproduct of brain activity and opens up new avenues for understanding the nature of reality and our place within it.

The Second Principle of Chaos Witchcraft: Many Worlds

MWI is a theory in quantum mechanics that suggests the existence of multiple parallel universes or realities. According to MWI, every quantum event results in the universe splitting into multiple branches or timelines, each corresponding to a different outcome. In other words, whenever a quantum system undergoes a measurement or interaction, the universe bifurcates into countless copies, each containing a version of reality where every possible outcome occurs. This interpretation implies that all possible alternative histories and futures are real and coexist within a vast multiverse.

Chaos Magick has always been about obtaining practical results. You can easily see this in action. Everyone knows someone whose life is constantly in a state of crisis and disaster. It seems as though every bad thing imaginable happens to this person, and yet they never seem to know why, and thus they live in victimhood consciousness, constantly generating undesirable synchronicity, and rarely take any responsibility for the state of their lives. The skilled Magician, on the other hand, utilizes this principle by altering their beliefs in order to generate desirable synchronicity and thus generates their own "luck."

The Third Principle of Chaos Witchcraft: String Theory

String theory, a theoretical framework in physics, proposes that the fundamental building blocks of the universe are not particles but rather infinitesimal vibrating strings. The vibrations of these strings create diverse manifestations in the physical world as they oscillate at different frequencies, giving rise to various particles and forces observed in nature.

String theory intersects with the many-worlds interpretation of quantum mechanics in intriguing ways. The notion of multiple parallel universes in MWI resonates with the concept of multiple dimensions or "brane worlds" proposed in certain versions of string theory. In this view, each universe or reality corresponds to a different configuration of strings and branes within a larger multiverse. From a magickal perspective, this interconnected web of realities could be seen as reflecting the interconnectedness and unity of all things and the infinite possibilities inherent in existence. It suggests that our perception of reality is just one facet of a vast cosmic tapestry, inviting contemplation of the nature of consciousness and existence itself.

When asked his opinion about theories concerning dark energy and dark matter,[2] Michio Kaku (one of the founders of string theory) said, "The best theory comes from string theory, which states that dark matter is nothing but a higher vibration of the string. We are, in some sense, the lowest octave of a vibrating string." This perfectly mirrors the principle of vibration found in *The Kybalion*, which states, "This principle explains that the differences between different states of matter, energy, mind, and even spirit, result largely from varying rates of vibration—the higher the vibration, the higher the position on the scale." Likewise, Chaos Witchcraft posits that Dark Energy, which is the expansive force in the universe, is pure consciousness, and the highest possible vibration.

Figure 15 illustrates how magick works from the Chaos Witchcraft point of view. As you move up the scale in connection and awareness, you move away from what Jung called "fate," which could be described as the most deterministic possible outcome for your life, and toward "destiny," which could be described as the most desirable outcome for your life. We cast spells, which generate desirable synchronicity, which act as bridges to higher and higher timelines and result in the manifestation of our desires.

The Fourth Dimension

String theory proposes the existence of extra spatial dimensions beyond the familiar three dimensions of length, width, and height.

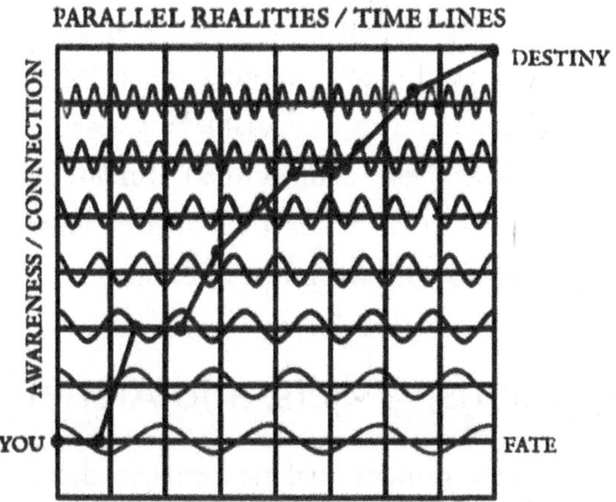

Figure 15. Scales of vibration

These additional dimensions, often depicted mathematically but not directly observable in our everyday experience, play a crucial role in string theory's attempt to unify the fundamental forces of nature.

From a mystical perspective, the idea of extra dimensions resonates with the concept of higher realms or planes of existence that are accessible through spiritual practice or altered states of consciousness. Just as string theory suggests that these extra dimensions are hidden from our direct perception, Chaos Witchcraft posits the existence of subtle realms or dimensions beyond the physical, accessible through inner exploration or transcendental experiences.

One way to interpret the fourth dimension within the context of string theory and mysticism is thus to consider it as a realm of expanded consciousness or higher vibrational states. In this view, the fourth dimension represents a level of existence beyond ordinary space-time, where consciousness transcends the limitations of linear time and physical form. Mystery traditions often describe experiences of timelessness, unity, and interconnectedness that parallel the concept of the fourth dimension in string theory.

String theory's mathematical framework suggests that these extra dimensions may be compactified or curled up into tiny, imperceptible shapes known as Calabi–Yau manifolds. From a mystical standpoint, these compactified dimensions could be hidden aspects of reality or dimensions of consciousness that are intricately intertwined with the fabric of existence.

Gods, Aliens, Angels, and Archetypes

In considering fourth-dimensional life forms, we delve into a realm where the boundaries between the physical and the metaphysical blur. Mythologies across cultures often depict gods and goddesses as beings possessing extraordinary powers, wisdom, and longevity, existing beyond the realm of mortal comprehension. Chaos Witchcraft posits that as inhabitants of higher dimensions—particularly the fourth dimension, as suggested by string theory—the gods exist in a very real way. Likewise, extraterrestrial life may in fact be *extradimensional* life.

It is endlessly intriguing and amusing to me that *The Book of Enoch* speaks of a classification of angels known as "The Watchers"—these beings, of course, would be fourth-dimensional—while we here in the third realm have created a two-dimensional reality called the internet which we are always staring at.

From this viewpoint, extraterrestrial civilizations could exist in dimensions where time behaves differently or where spatial coordinates are configured in ways beyond our comprehension. They may navigate realities governed by different physical laws or possess abilities and technologies that go beyond our current understanding of science and technology.

From the perspective of Chaos Witchcraft, the gods and goddesses of mythology transcend our three-dimensional reality and exist within a realm of expanded consciousness and heightened vibrational frequencies. Just as string theory proposes that extra dimensions are hidden from direct observation, the gods may dwell in a dimension beyond the boundaries of time and space. Jung also posited what he called "acausal connecting principles" that operated outside of time and space, existing in realms inaccessible to mortal perception.

The gods of mythology often interacted with humanity, shaping destinies, imparting wisdom, and influencing the course of events. Viewing them as fourth-dimensional beings creates a framework to understand their ability to traverse time and space, manifest in various forms, and wield profound influence over the fabric of reality.

The concept of gods as fourth-dimensional entities not only aligns with traditions that posit the existence of higher planes of consciousness and spiritual dimensions, but also aligns perfectly with the concepts of string theory and MWI. Just as mystics seek to transcend the limitations of the physical realm and commune with divine forces, the gods of mythology may be manifestations of transcendent consciousness operating at higher dimensional levels.

The Thought Bomb

For this exercise we will be visualizing dark energy, the force responsible for the expansion of the universe, as a means to penetrate dark matter and thus bring forth a new possibility or timeline bridge—consciousness being the primordial substance that shapes reality, and dark energy being its highest vibration.

At your altar or ritual chamber, begin by casting the Chaos Circle (Figure 16) in your mind. Invoke the fourth-dimensional elements of time, space, energy, and entropy by stating in a loud, firm voice, "I invoke the elements of the fourth dimension into this space that I may command them to do my bidding."

Figure 16. Chaos Circle thought bomb

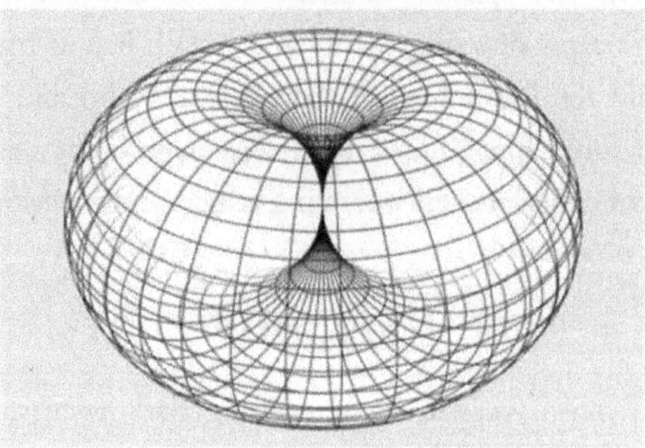

Figure 17. The spindle torus of energy

Now visualize a toroidal sphere of energy (Figure 17) surrounding you, spinning faster and faster, becoming both a magnet for your desires and a force capable of penetrating all of reality. Focus all of your will and intention into this sphere of dark energy that now surrounds you, giving off an incredibly bright light. Your intention can be anything, any change you want to see in your exterior reality—but please keep in mind the central equation of magick. Manifesting a Ferrari into your garage is unlikely, but manifesting a promotion or business opportunity that leads you to the Ferrari is not at all out of the realm of possibility, so long as you are willing to act when divine inspiration demands.

Gather the maximum amount of focused mental energy into your sphere that you are capable of, and when you cannot possibly contain this energy any longer, allow it to explode outward. Visualize this like a bomb of exploding mental and emotional energy, at the center of which is you.

This is the thought bomb—a powerful way to push conscious thought into the aether and impregnate it with your will.

Conclusion

Chaos Witchcraft may be a far cry from the Chaos Magick of the great Peter J. Carroll, but it is true to Chaos Magick's core axioms: the use of belief as a tool, and the search for the underlying metaphysical structure of magickal phenomena. It is a sincere attempt to explain from a scientific point of view that which has been unexplainable by mainstream science, and yet experienced by every sincere practitioner of magick since our ancestors first carved mysterious symbols and sigils on cave walls . . . and at the end of the day, it's only a belief system.

15

The Mythogenesis of Baphomet

BY JOZEF KARIKA

I

Some have misguided ideas about Baphomet. They think of Her/Him as a benevolent and caring Mother Nature concerned for their well-being. This is true in a way, but not in the manner these Gaia enthusiasts imagine. In this myth we are creating, Baphomet (aka Thanateros aka Biosphere) originates from the Outer Gods; She/He is their progeny and echoes their characteristics. Not so nice, at least from a human point of view, if you're familiar with Lovecraftian lore. Arthur Schopenhauer, too, with his vision of the ravenous, all-devouring Will, hit it right. Also Castaneda with his Eagle, which after death tears and devours the remains of human consciousness with its beak.

First, we need insight into Baphomet in a wider context. This includes Chaos (aka Azathoth). The short version reads like this:[1]

> *However we choose to see it [Chaos], the ultimate ground of being is utterly void to our understanding, impersonal and inhuman, whimsical and capricious and far too infinite and incomprehensible to be much use as a god to limited dualistic beings like ourselves.*
>
> *There is a part of Chaos which is of more direct relevance to the magician. This is the spirit of the life energy of our planet. All living beings have some extra quality in them which separates them from inorganic matter. The ancient shamans mainly sought to represent this force by the Horned God. In more modern times, this force has reasserted itself in our awareness under the symbol of Baphomet. (Carroll 2022, 140)*

For a slightly more detailed version, we call on Schopenhauer. Although he did not write about Chaos and Baphomet, but about the Will (*der Wille*), which we may or may not agree with, his insight becomes quite illuminating, even if we see it only as an illustrative model. It goes something like this (very simplified): Schopenhauer conceived the Will as an all-pervasive, metaphysical force that underlies everything in the universe. He considered it to be an irrational, blind, and instinctive energy, devoid of consciousness or higher purpose. The Will manifests in nature even before the emergence of life. Before life appears, the Will operates in what he calls the "inorganic" realm of nature. Here the Will is evident in gravity, electromagnetism, and other fundamental forces that govern the behavior of matter and energy. These forces exhibit an inherent drive for self-preservation and a tendency to seek equilibrium. In this context, the Will is an underlying, non-conscious force that drives the natural processes and phenomena that shape the universe.

Schopenhauer believed that the Will's manifestations in the inorganic world are not guided by reason or conscious intent, but follow a blind, deterministic impulse toward maintaining the order and structure of the universe. It is through these natural forces that the groundwork for the emergence of life and organic beings is established. The Will's boundless hunger leads to perpetual strife and conflict, both within individuals and in the external world. On the individual level, competing desires give rise to internal battles, leaving individuals torn between conflicting impulses (sub-selves). Externally, Schopenhauer saw the world as a battleground where various entities, driven by their respective Wills, vie for resources, dominance, and survival, with the animal (and human) body and all its parts being the visible expression of the Will and its several desires. The teeth, throat, and bowels, for example, are "objectified" hunger. He argued that our desires, emotions, and aspirations are all also manifestations of the underlying Will. Even consciousness, reason, and intellect, in his view, just serve the purpose of assisting the Will in attaining its goals, rather than being independent guiding forces.

Thus, mythologically, the Will (Chaos, Azathoth) created life as such in the cosmos (Shub-Niggurath), and this Outer God then spread the seed of life on billions of planets through countless cosmic eons, including Earth. This terrestrial life (Baphomet) further evolved itself by later creating consciousness, senses, and finally intellect to better satisfy its needs. All the categories through which we grasp and filter reality, all the temporal and spatial forms (Yog-Sothoth), are also just creations of the Will, which we use to navigate reality while satisfying its desires. The Will lies "behind" these forms, so we cannot say or think anything meaningful about it.[2]

Note that Schopenhauer's characterization of the Will and its operation in the universe, even regarding its "creation," corresponds with Lovecraft's and Carroll's view of Azathoth.

In this mythos, we call that segment of the Will/Chaos/Azathoth that we can perceive and conceptualize as terrestrial life (and perhaps proto-life in inorganic nature) by the name of Baphomet.[3]

II

In his book *The Doctrine of Awakening*, Julius Evola vividly describes the experience of Baphometian consciousness, which he refers to as saṁsāric consciousness:

> *There are some who, at certain moments, are able to become detached from themselves, get beneath the surface, down into the dark depths of the force that rules their body, and where this force loses name and identity. They have the sensation of this force expanding and including "I" and "not-I," pervading all nature, substantiating time, supporting myriad beings as if they were drunk or hallucinated, reestablishing itself in a thousand forms, irresistible, untamed, inexhaustible, ceaseless, limitless, burning with eternal insufficiency and hunger. He who reaches this fearful perception, like an abyss suddenly opening, grasps the mystery of saṁsāra and of saṁsāric consciousness and understands and fully lives anattā, the doctrine of nonaseity, of "not-I." (Evola 1996, 88–89)*

My experience of the full Baphometian invocation was slightly different. I did it over winter high up in the mountains, at a location I've intuitively associated with Baphomet since my first visit.[4] The mountains were without people, with a presence of wild animals, which added to the sense of danger. This impression, along with the rhythmic sound of the snow breaking through as I walked, created a trancelike state (gnosis) even while arriving at the site. I performed the Gnostic Octaris Ritual (GCR), then the Mass of Chaos, strongly visualized Baphomet in front of me, and then merged with Him/Her. It was powerful, almost psychedelic, although I didn't take any elixirs, drugs, or alcohol. I felt as if I had fallen into the depths vertically below me to the core of life, but also to "proto-life" in rocks or metals.

Along with this fall, my perception of self fragmented into countless tiny selves, perhaps thousands, millions, constantly crumbling into smaller and smaller units (each with its own increasingly tiny "me"), until they merged into a single shapeless, swirling, bubbling mass. This mass felt kind of dull, blind-willing, but was otherwise dark, without our kind of intelligence. I felt only its constant urge to generate, to create, to feed, to devour. It had no regard for the individual forms that it was constantly spawning and absorbing; it probably didn't even perceive them, in the same way that we don't perceive the individual cells in our bodies. I (we—thousands) was part of it. Certainly there was no individuality or ego in that experience, only a vague multi-awareness that somewhere "up there" on the surface there was an absurdly transient, unstable, and trivial formation with my personal name; billions of similar ones were emerging from this mass and billions of them would perish in the same way.

On the one hand it was extremely discouraging, on the other equally fascinating. I felt the impulse to immerse in this mass and its formations and to participate fully, almost ecstatically, in them, to ride their wave and diversity indefinitely. At the same time, I also felt the opposite impulse to detach myself from this current, to step out of it into nothingness and stillness. During this experience, my perception of time blurred. However, when I returned, performed the banishing, and grounded myself in a normal state of consciousness, the strong impression of these two opposing impulses remained.

III

So far, so good. What's the problem? My experience suggested what can also be discovered by careful observation. Carroll describes how this insight became evident to Charles Darwin, who was one of the first to recognize it:

> *Darwin himself had at first considered becoming an Anglican priest; however his eventual insights into natural history proved devastating. Species, including ours, had not leapt fully formed*

> *into creation as biblical accounts imply, but had evolved over astonishingly long periods, as geology could confirm. Christian theologians could in many cases reconcile themselves to accepting that god had used an evolutionary mechanism and simply failed to mention it in the bible for the sake of simplicity, but few could accept the implications of the horrifying nature of the underlying mechanism. Of all animals ever conceived, only a tiny fraction ever live out their full lifespans; the rest succumb to premature death from predation, starvation, internecine violence, parasitism and disease.*
>
> *Natural selection acts with astonishing harshness upon randomly generated recombinations and mutations within gene pools. The whole process appears stochastic rather than teleological, driven by harsh and bloody chance rather than any divine purpose. Despite appearances, nature now looked very far from "All Things Bright and Beautiful" made by the benevolent deity of the English country clergyman. It looked more like something cobbled together by a less than omnipotent Demiurge or Archon with a perverse and vicious sense of humour. (Carroll 2014, 7)*

Baphomet improves itself through antifragility, which is its defining principle. However, as Nassim Taleb writes:

> *There is a different, stronger variety of antifragility linked to evolution that is beyond hormesis[5]—actually very different from hormesis; it is even its opposite. It can be described as hormesis—getting stronger under harm—if we look from the outside, not from the inside. This other variety of antifragility is evolutionary, and operates at the informational level—genes are information. Unlike with hormesis, the unit does not get stronger in response to stress; it dies. But it accomplishes a transfer of benefits; other units survive—and those that survive have attributes that improve the collective of units, leading to modifications commonly assigned the vague term "evolution." . . . So the antifragility of concern here is not so much that of the organisms, inherently weak, but rather that of their genetic code, which can survive them. The code doesn't really care about the welfare of the unit itself—quite the contrary, since it destroys*

> *many things around it. . . . In fact, the most interesting aspect of evolution is that it only works because of its antifragility; it is in love with stressors, randomness, uncertainty, and disorder—while individual organisms are relatively fragile, the gene pool takes advantage of shocks to enhance its fitness. So from this we can see that there is a tension between nature and individual organisms.*[6] *(Taleb 2012, 66–67)*

But before you start gloating about Baphomet's gift of antifragility (your organic heritage), you should read the fine print. You will see that your idea of antifragility and Baphomet's are profoundly different:

> *So, in a way, while hormesis corresponds to situations by which the individual organism benefits from direct harm to itself, evolution occurs when harm makes the individual organism perish and the benefits are transferred to others, the surviving ones, and future generations. For an illustration of how families of organisms like harm in order to evolve (again, up to a point), though not the organisms themselves, consider the phenomenon of antibiotic resistance. The harder you try to harm bacteria, the stronger the survivors will be—unless you can manage to eradicate them completely. (Taleb 2012, 70)*

Hence the hard truth about Baphomet as a god in two points:

1. He/She doesn't see you as an individual at all, just like you don't perceive your individual cells or genes. His/Her "perception," if any, starts on a much larger taxonomic scale of species, genus, etc. For humans, perhaps at the minimal level of tribe or race or somewhere in between. Similarly, you perceive a cluster of cells only from a certain size. Therefore, it is useless to invoke Baphomet as a god in any individual matters except to stimulate the source of all your drives, the vital basal layer underlying your multi-self.

2. Even if He/She could perceive you, He/She doesn't care about you as an individual.[7] And He/She certainly doesn't care about your personal happiness, health, success, achievements, etc. If anything, you serve Him/Her by your interesting and untimely death, because He/She will extract from it the most valuable feedback, micro-information that He/She will (blindly) incorporate into the next generations of humans. An even better service you can render Him/Her is to beget as many offspring as possible and then die. He/She won't thank you or reward you with any nice afterlife.

IV

This is, I admit, an extremely dark view. However, this reality can also be seen from another, much more joyful and affirming perspective. Whatever may strike you as an individual (or all of humanity) as evil or even devastating is just a lesser degree of good for Baphomet.

Compared to Schopenhauer, Nietzsche seems more optimistic in that you as an individual can foster Life not only by your death, but also by your self-overcoming during your lifetime. In his vision, it is this self-overcoming that is the defining characteristic of Life—which is confirmed by the various stages in the symbol of Baphomet, always creating something over and above itself.[8] Baphomet evolves from proto-life through primordial life forms, from animal stages to man with intellect, creativity, and magic, potentially to an interstellar state—this expansion being perhaps His/Her primal meta-instinct. As an individual, you can contribute to this by amplifying what you have in common with Baphomet: that primal meta-instinct, i.e., the will to power.[9]

Nietzsche perceived Baphomet's nature (calling Him/Her "a Life") more accurately than Schopenhauer. He was probably the first philosopher to capture antifragility, as well as the notion that Baphomet seems characterized not only by a mere will to live (Schopenhauer's vision)—that is, driven by self-preservation, survival, and the satisfaction of desires—but by a will to power (*Wille zur Macht*)—that is, the drive to grow, expand, become more powerful, capable, effective, creative. In other words, we are motivated by not only survival but transformation, the overcoming of our own forms and the endless creation (and destruction) of new ones,[10] even at the cost of risk, and even sometimes sacrificing individual self-preservation.[11]

In his book *The Affirmation of Life*, Bernard Reginster rigorously analyzes Nietzsche's writings and comes to this conclusion:

> *The will to power is therefore the will to "striving against something that resists." Since striving against is an effort to overcome, we might say that the will to power is the will to overcoming resistance. . . . Nietzsche explicitly distinguishes the will to power from the will to happiness. This suggests that the resistance to overcome is resistance against the satisfaction of desires and that the will to power is not a will to the state in which resistance has been overcome (a state in which desires have been satisfied), which is "happiness" in the sense presupposed by Schopenhauer's pessimism. Furthermore, the will to power is not simply a will to resistance, the desire for a condition in which some determinate desire is perpetually frustrated by resistance or obstacles to its fulfillment. There would be no "expansion, incorporation, growth" unless the striving was eventually successful. The will to power, in the last analysis, is a will to the very activity of overcoming resistance—"the will's forward thrust and again and again becoming master over that which stands in its way," or "the game of resistance and victory," which consists of "a little hindrance that is overcome and immediately followed by another little hindrance that is again overcome."* (Reginster 2006, 126–127)

In this sense, then, the will to power, the "essence" of life (and of Baphomet), paradoxically "desires displeasure."[12] After a careful examination of the relation of the will to power to other desires and drives in the perspectives of several philosophers interested in Nietzsche, Reginster comes to this conclusion:

> *In my view, then, the will to power is the will to the overcoming of resistance. This definition dictates a particular conception of the relation between it and other drives. So defined, the concept of power is, in and of itself, devoid of any determinate content. It gets a determinate content only from its relation to some determinate desire or drive. . . . Accordingly, the will to power cannot be satisfied unless the agent has a desire for something else than power. . . . The will to power therefore has the structure of a second-order desire: it is a desire whose object includes another (first-order) desire. It is, specifically, a desire for the overcoming of resistance in the pursuit of some determinate first-order desire. (Reginster 2006, 131–132)*

Now we get to the point: the Octaris or Chaostar as a symbol of the will to power and also a visual blueprint for our working as Chaos Magicians. The Octaris as a whole (no particular part of it) symbolizes the will to power, the second-order desire to expand, to overcome resistance, to grow. Baphomet as a hidden god underlies the eight major organic desires that spring from Him/Her (in human biology). These first-order desires delineate eight vectors and specific goals in the direction of which the Chaos Magician exercises the will to power; i.e., overcomes resistance to manifesting or achieving them.[13] (The contemporary magician, however, no longer fulfills simply the eight major organic desires, but the synthesis of these desires with the corresponding cultural meme complexes. This produces a personal pantheon consisting of the eight major deities.)

For now, this roadmap will suffice.[14] The magician, by five kinds of conjurations as an addition to the mundane methods, perpetually achieves, grows, overcomes resistance to the attainment of goals along the vectors of these eight desires/drives/quests/games.[15] In doing so, they continuously overcome their eight ego-stories, ego-concepts. On each vector lie specific, well-defined short-, medium-, and long-term goals as checkpoints important for manifesting that vector in reality. These particular goals/objectives matter, as without them, you're in the business of LARPing or daydreaming, not magic. Yet they do not constitute the crux of the Chaos Magical activity, which is an endless and fractal process of achieving in given vectors or directions.[16]

Here I probably slightly differ from the view of the archmage and my mentor, Peter Carroll. In my vision, the primary desire of the Kia incarnation is not just to taste as much freedom and variability of experience as possible. These form a vital aspect of the earthly incarnation, but in the magician's life they should not exist without the organizing principle, which is the will to power—the constant overcoming of resistance along the eight vectors.

I see two extremes here to avoid: Scylla and Charybdis. On the one side, the obsession with a particular life task and a rigid ego-concept. Closure from Chaos and randomness, succumbing to one's own illusion; what we call the demon Choronzon. On the other side, aimless drifting and tasting of experiences and sensations, complete succumbing to Chaos and randomness without any inherent organizing principle. Moreover, a life without continual overcoming of resistance leads not to satisfaction and joy but to dissatisfaction and disintegration, regardless of how colorful and variable it may be in terms of external stimuli.[17]

In this context, I present a baseline roadmap:[18]

Mercurial drive, orange magic: Quest for intelligence, knowing, thinking, understanding, proper use of the mind, satisfying curiosity, commerce, entrepreneurship, short-term trading, speculation. Also for writing, communicating, marketing, selling, negotiating, persuading, decision making, luck management.

Venusian drive, green magic: Quest for relationships (romantic, friends, business, social); their establishing, maintaining, cultivating. For social intelligence, connections with people, nature, the world. Also for beauty, artistic expression/enrichment, elegance, fashion, aesthetics, love, loyalty, harmony, relaxation, fun, leisure, joy, pleasure, recreation, empathy, mindfulness, sensuality, benevolence.

Jupiterian drive, blue magic: Quest for wealth, a rich portfolio of assets and their prudent management, long-term investing, economic sovereignty and/or political power, leverage. Acquirement, preservation, and enhancement of f*ck you money. Also for joviality, overall well-being, natural authority, reputation (even rulership) in the personal, professional, and social spheres.

Solar drive, yellow magic: Quest for individuation, freedom, autonomy, sovereignty, independence, self-esteem, rational self-interest, intentional living, productive work, indistractability, attention control, ability to focus, flow activities, charisma. Also for health (physical/mental), bodymind integrity, and resulting organic joy.

Saturnian drive, black magic: Quest for acceptance of mortality (own and of close ones), for motivation and intensity drawn from the awareness of finality, the scarcity of time, the uncertainty of life. Also for clarity, discipline, efficiency, perseverance, simplicity, resilience, asceticism, longevity, robustness, understanding the cycles of life, and dying well.

Lunar drive, purple/silver magic: Quest for creativity, imagination, intuition, willingness to feel and experience, sensitivity, nurturing, sexuality, sexual kinks. Also for all kinds of procreation, whether artistic (paintings, sculptures, books, music, films, etc.) and/or begetting children. All matters of family and family relations.

Martial drive, red magic: Quest for vigor, physique, strength, vitality, resilience, fitness, adventure, courage, survival and fighting skills, assertiveness, stamina, zest for life, toughness (even ruthlessness), harnessing the power of anger and rage, overcoming fear. Also for aggression in the sense of movement toward (not away from) relationships, persons, situations, experiences, reality (the opposite of regression, withdrawal from reality, retreat into illusion).

Ouranian drive, octarine magic: Quest for magical power and knowledge, edge between genius and madman, life's work, awareness and integration of all eight sub-selves into your unique vibrant pattern. Also for all things strange, antinomian, weird, aetheric (Carroll), daimonic (Harpur), superspectrum (Keel), hidden, belonging to the other/outer side. In this digital age, also the art of invisibility and privacy.

As a Chaos Magician you need to pay attention to all eight of these major drives (and sub-selves formed around them).[19] As we already know, everything in the bodymind stays interconnected, so a long-term frustration of any drive will sooner or later disrupt the homeostasis of the whole. The same goes for oversatisfying one drive (sub-self) at the expense of others, or satisfying them in a ratio that is unnatural to you. The pattern and ratio of these drives seem unique in each human (your gift from Baphomet).[20] Moreover, they continuously fluctuate depending on the life context, life stage, and way of interaction with the environment. Therefore, working with all eight is vital. The Chaos Magician should achieve and maintain at least a moderate state of satisfaction in all eight, which requires constantly overcoming resistance, as I described above. At least with three or four drives, however, the magician should aim much higher, desiring much more than the usual civilian satisfaction and manifestation. That's the purpose of magic, especially of the traditions oriented toward life. The trick is achieving such satisfaction through continuous dissatisfaction, while not shattering the homeostasis of the whole.

A simple method of such working looks like this. Every three months, sit down and number the importance of each drive (subself) to you in your current life stage and situation, from one to ten. (One means minimally important, ten means maximally important.) Then repeat the process, but this time number your satisfaction with how much attention and specific real-world actions you devoted to each drive over the past two weeks.

Pay the most attention to the drives for which you see the widest gap between their actual importance to you and your satisfaction with fulfilling them. For such drives, devise concrete goals and execute specific flow activities, habit-forming activities, and conjurations of all five kinds[21] to achieve these goals. Pursue these activities with commitment and perform checks on a daily, weekly, and monthly schedule. After three months, repeat the process, then again, for the rest of your life.

The ongoing appeal and variability of this meta-quest or metagame (composed of eight quests/games) stems from the following factors:

1. The more effective and experienced you become in each game/quest, the higher, greater, more ambitious challenges to overcome you see. After climbing a mountain and savoring a brief satisfaction, you immediately start thinking of the next, even higher one. This applies even after you've achieved mastery in a given game. The skill of pursuing a quest can be refined until death, or as

long as your physical and mental condition realistically permit. However, this doesn't mean that you can never feel content, or that what you have achieved is enough. Neither does it mean that you have to keep increasing the risks, demands, and strain in every game until you break, come to ruin, burn out, or jeopardize or blow up the other games. Each of the eight major games has dozens of sub-games, providing dozens of different objectives and varying experiences. You can (and should) switch between these sub-games according to your changing preferences, external circumstances, and different life phases.[22] Not too often, though, so you don't jump between the initial (easy) levels of each sub-game or change sub-games every time they become demanding or a setback occurs. Continuous achieving and overcoming have to remain present.

2. As you get older, you gain experience, but your physical and mental abilities slowly decline. Hence, even to maintain the status quo, you will have to increase your efficiency and commitment. The game gets harder and harder as time goes on; all meta-quests eventually end tragically. But therein lies some grim amusement—this gives intensity to your meta-quests.

3. Unexpected and random setbacks in life can push you backward or to square one in a given game (or even into the negative range in some games, such as finance or health).[23] Make or break, swim or sink.

Through the pursuit of this eightfold achieving, taking into account your constantly changing environment, life context, and stage, your unique Octaris will give meaning to your existence. This your Kia willed/desired when incarnated, objectified in your human body—flesh and bones.[24] Working with the unique inner (organic) Octaris becomes even more important in the current Pandaemonaeon, when all external supports and directives in the form of traditional belief systems, collective norms, politics, or institutions have fallen away. Now nothing is true; everything is permitted. As a consequence, with the help of twenty-first century technology, all the demons are unleashed, battling in both physical and virtual space for your attention, time, sanity, energy, belief, flesh, and brains. Without your Octaris to guide you through this fluid, volatile environment, you might easily become lost, distracted, hooked, harvested, and devoured.

ABOUT THE CONTRIBUTORS

Carl Abrahamsson is the author of occultural books such as *Reasonances* (2014), *Occulture: The Unseen Forces That Drive Culture Forward* (2018), *Anton LaVey and the Church of Satan* (2022), *Source Magic* (2023), and *Meetings with Remarkable Magicians* (2024). He is the editor and publisher of the magico-anthropological journal *The Fenris Wolf*, and also occasionally makes films and music. He lives in Vimmerby, Sweden, together with his wife, American psychoanalyst and artist Vanessa Sinclair.

Peter J Carroll has developed the ideas and practices of Chaos Magic since the mid-1970s, and for a number of years he led the Magical Pact of the Illuminates of Thanateros, which helped to disseminate Chaos Magic worldwide. He wrote the seminal Chaos Magic texts *Liber Null & Psychonaut* (1987) and *Liber Kaos* (1992) and has since written another five books on magic and alternative physics. Since retiring from a successful career in business he has devoted his time to his family and esoteric research, and to interacting with aspiring magicians worldwide via his Arcanorium College. He lives in Southwest England.

Ivy Corvus has been practicing various occult philosophies for over 15 years. She is best known as "Ivy The Occultist" on YouTube, where she has cultivated a reputation as an experimental Chaos Magician. Through her quickly growing channel, established in 2022, Ivy shares occult knowledge, introduces innovative concepts in Chaos Magic, conducts interviews with esteemed occult authors and practitioners, and fosters new ideas that represent a new generation within the occult community.

Ivy has always expressed an interest in the intersection of science and spirituality. She is both a healthcare practitioner and an undergraduate student, double majoring in psychology and molecular biosciences. Her passion lies in understanding the interconnectedness of the universe and guiding others along their spiritual journeys. She lives in the Pacific Northwest with her partner, her animals, and her garden.

Sanhre Daffowt is a passionate seeker of metaphysical truths and occult wisdom, with a profound fascination for the intersection of science and magick. Initially captivated by LaVeyan Satanism, Sanhre's quest evolved into a profound exploration of Chaos Magick, Hermetic philosophy, astrology, and Zen Buddhism. Delving into the profound potential of belief, Sanhre integrates the principles of the many-worlds interpretation of quantum physics (MWI) with ancient occult traditions, forging a unique path as a modern-day Chaos Witch. Embracing belief exploration as a powerful tool for personal empowerment, Sanhre advocates for the transformative potential of Chaos Magick and the Occult Sciences, inspiring others to embark on their own journeys of self-discovery and growth through esoteric wisdom.

Sanhre is the host of *The Order of Chaos* podcast, which has featured many of today's most prominent and influential occultists from around the world, as well as the founder of The Order of Chaos Mystery School.

Jaq D Hawkins is the author of *Understanding Chaos Magic* (1996; now absorbed into *The Chaonomicon*), *Chaos Monkey* (2002), and *Elemental Spirits* (2024), a single volume comprised of her *Spirits of the Elements* series, as well as several fiction titles. She has lectured widely on various occult topics at pagan and occult events and is a former lecturer at Arcanorium College. She lives in Buckinghamshire, England, with her family, which includes several cats, and works on amateur filmmaking when she isn't writing. Most of the time she juggles deadlines between the two.

Information on Mind, Body, Spirit titles can be found at *www.jaqdhawkins.co.uk*.

Ronald Hutton is Senior Professor of History at the University of Bristol, the Gresham Professor of Divinity at Gresham College, London, and a Fellow of the Royal Historical Society, the Society of Antiquaries, the Learned Society of Wales, and the British Academy. He has held a number of public posts and is currently a member of the Historic Estate Conservation Committee of Historic England and also the chair of the first national Blue Plaques Panel. He has published eighteen books and ninety-five essays on a wide range of subjects including British history between 1400 and 1700, ancient and modern paganism in Britain, the British ritual year, and Siberian shamanism.

Jozef Karika is a Central European author of twenty-two published books, of which more than six hundred thousand copies have been sold. Two of his novels have been adapted into successful cinema releases, and one was also made into a TV miniseries. He has been applying the techniques of Chaos Magic to enhance his creativity, intentions, and imagination since 2006, when he attended Peter Carroll's courses at Maybe Logic Academy and later at Arcanorium College, where he attained a Bachelor of

Magic degree. In the non-magical academic world, he has master's degrees in history and philosophy.

Sinobu Kurono began studying Golden Dawn magic at the age of eight and initiated his training in Chaos Magic at sixteen (over thirty years ago). He was the first to introduce authentic Chaos Magic to Japan and served as the head of the Japanese branch of the Illuminates of Thanateros (IOT) from 2000 to 2002; however, due to internal conflicts and conspiracies within the Japanese branch, he was subsequently expelled from the IOT. Currently, he is an educator in the Order of Chaos Knights, an organization that has been active for over twenty years, where he utilizes experience gained while studying at Arcanorium College. He has authored three books on subculture, four books on Chaos Magic, and one novel, and is the sole author of Chaos Magic books published commercially in Japan.

Sinobu served as the resident DJ at psychedelic trance parties for a decade and continues to work as a DJ, as well as handling vocals and synthesizer duties for a band. Additionally, his Chaos Ring has gained recognition internationally, and he and his colleagues are renowned for creating the world's most beautiful unbreakable Chaosphere. While it comes with a high price, it is crafted using materials such as fiber-reinforced plastic (FRP), used in artificial satellites, and paint from Japan's luxury cars. In strength tests, the sphere remains intact even when dropped from a height of three meters.

Dave Lee's magical practice is rooted in teenage psychedelia but became rather more focused when the multi-model approach known as Chaos Magic came along. He was a founder member in 1980 of the first ever working group of the Illuminates of Thanateros, and still serves in that community as an Elder. This work has given him an enormous amount of experience in facilitating group ritual, and since the lockdowns he has been working with a number of magical groups to develop techniques and approaches for working group magic online. All his public-facing links are at *linktr.ee/david23lee*, where readers can also sign up for his free events and newsletter.

Dave lives in Sheffield, UK, with his favorite human being, a collie, and two cats.

Mariana Pinzón, aka Soror Mavis, is a Berlin-based transformational coach, ritual designer, Discordian Witch, and Psychedelicious Nun of Chaos at the O:D:D. She is the founder of the ChaoSurfing project and a member of the German section of the IOT and the Psychedelic Society Berlin.

In her work, she supports edge-dwellers moving through chaotic life transitions to regain balance and craft a new path. She also offers ritual workshops, magical quests, online courses on Chaos Magick and the Eight-Circuit Brain model, psychedelic integration sessions and individualized ritual experiences, and grief support, as well as support with shadow work.

She holds a Magister Artium degree from the University of Heidelberg in Comparative Religion, Philosophy, and Political Science and an ICF accredited coaching diploma from the Animas Center for Coaching.

For more information on her work, visit *chaosurfing.rocks*.

Jacob Sipes, PhD(c), BoM, is an academic philosopher, award-winning filmmaker, critically acclaimed musician, published author, martial artist, world explorer, and mountaineer. He is a PhD candidate at Warwick University, where his thesis examines psychological processes and their implications for epistemological methods.

Jacob was the first to complete the bachelor's program at Arcanorium College, where he wrote a bachelor's thesis investigating magic fundamentals. He now specializes in meta-magic research, which seeks to describe the shared principles, components, and features that make disparate magical practices successful. His current research investigates how we might describe and better use the natural phenomena typically employed by a magic practice.

His current magic project is *The Septhed: Gods of Kenilworth*, a theurgical art project that spans literary, musical, and film mediums. His latest installation, titled "The Lover," is a love song and music video that serves as a group invocation of the Aphrodite principle. The song and video, premiered at Slipknot's Knotfest in 2023, earned a front-page feature on the world's most popular rock magazine, *Digital Noise*, where the full-page review can be found.

Jacob and his wife currently reside in England, in the West Midlands. However, they are often not found there, but gone on some adventure abroad. They celebrated 10 years of marriage in February 2024.

Lionel Snell has been interested in magic for over 70 years. While reading pure mathematics at Cambridge University, he studied the works of Crowley and later Austin Spare. In 1972, his article "Spare Parts" was the first published account of the magical system described in Spare's *The Book of Pleasure*. This—together with his books *SSOTBME: An Essay on Magic* and later *Thundersqueak*, published under the pen name Ramsey Dukes—became a key influence on the nascent Chaos current.

Lionel was initiated into the IOT in the 1980s and later into the Caliphate OTO, where he was the senior initiate in South Africa for more than a decade. He has published a dozen books on magic and was described in a recent *Weird Studies* podcast as follows: "Ramsey Dukes, also known by his real name of Lionel Snell, may be one of the most important thinkers on magic since Aleister Crowley."

Hagen von Tulien is a contemporary artist and occultist. With about forty years of experience in intense magical theory and practice, he has specialized in creating art as an expression and manifestation of magical states of awareness and its use as an esoteric tool. He works in a variety of media, including pen and ink, paper cut, collage, and digital formats. During the 1980s he came in contact with the current of Chaos Magic, which has never ceased to attract and fascinate him since. Throughout the 1990s Hagen was a key figure in the Magical Pact of the Illuminates of Thanateros (IOT), serving as its section head for Germany. Besides Chaos Magic, he also devotes his time to the study and practice of many other occult systems, approaches, and ways.

He also is a Gnostic Bishop of the Ecclesia Gnostica Spiritualis (ordained by Michael Bertiaux), a Master-Initiate of the Fraternitas Saturni (FS), and an empowered adept of the Voudon Gnostic Current (member of OTOA/LCN), focused on deeply researching the Gnosis of the Saturnian Continuum and of Esoteric Voudon.

Julian Vayne is a British occultist, independent scholar, and author with over four decades of experience within esoteric culture: from Druidry to Chaos Magic, from indigenous shamanism through to Freemasonry and witchcraft. Julian is a co-organizer of the psychedelic conference Breaking Convention and a founding member of the post-prohibition think tank Transform. He sits on the academic board of *The Journal of Psychedelic Studies* and has been a visiting lecturer at several British universities. Julian facilitates psychedelic ceremony, as well as providing one-to-one psychedelic integration sessions and support. In addition to exploring traditional sacred medicines, Julian's work includes the first published accounts of the entheogenic ritual use of ketamine and several novel psychedelic sacraments.

Aidan Wachter is an animist dirt sorcerer, artist, jeweler, musician, and author. His work reflects a wide-ranging eclectic and non-dogmatic approach with influences from many paths. He believes that magic is an innate human talent that can be developed with intentional practice, rather than something that must be granted to the seeker by outside forces, be they human or Other. He is the author of *Six Ways: Approaches & Entries for Practical Magic* (2018), *Weaving Fate: Hypersigils, Changing the Past, and Telling True Lies* (2020), and *Changeling: A Book of Qualities* (2021).

REFERENCES

Chapter 3

Assagioli, R. 1965. *Psychosynthesis*. Hobbs, Dorman & Co.

Branden, N. 1994. *The Six Pillars of Self-Esteem*. Bantam Books.

Carroll, P. 1987. *Liber Null & Psychonaut*. Red Wheel/Weiser.

Carroll, P. 1992. *Liber Kaos*. Red Wheel/Weiser.

Carroll, P. 2010. *The Octavo*. Mandrake of Oxford.

Carroll, P. 2022. E-mail correspondence.

Csikszentmihalyi, M. 1990. *Flow: The Psychology of Optimal Experience*. Harper & Row.

Csikszentmihalyi, M. 1996. *Creativity: The Psychology of Discovery and Invention*. HarperCollins.

Duke, A. 2019. *Thinking in Bets*. Penguin.

Gendlin, E.T. 1986. *Let Your Body Interpret Your Dreams*. Chiron.

Goleman, D. 2013. *Focus: The Hidden Driver of Excellence*. Harper.

Greene, R. 2018. *The Laws of Human Nature*. Profile Books.

Hayes, S.C. 2019. *A Liberated Mind*. Avery.

Johnson, R. 1986. *Inner Work—Using Dreams and Active Imagination for Personal Growth*. Harpercollins.

Johnson, R. 1991. *Owning Your Own Shadow: Understanding the Dark Side of the Psyche*. HarperCollins.

Kaufmann, W. 1950. *Nietzsche: Philosopher, Psychologist, Antichrist*. Princeton University Press.

Lachman, G. 2004. *In Search of P.D. Ouspensky: The Genius in the Shadow of Gurdjieff*. Quest Books.

Nietzsche, F. 1982. *The Portable Nietzsche*. Edited and translated by Walter Kaufmann. Penguin.

Ouspensky, P.D. 1957. *The Fourth Way*. Alfred A. Knopf.

Parrish, S. n.d.a. "Iatrogenics: Why Intervention Often Leads to Worse Outcomes." *Farnam Street blog. fs.blog*

Parrish, S. n.d.b. "Second-Order Thinking: What Smart People Use to Outperform." *Farnam Street blog. fs.blog*

Seneca, L.A. 1969. *Letters from a Stoic*. Translated by Robin Campbell. Penguin Classics.

Taleb, N. 2012. *Antifragile: Things That Gain from Disorder*. Random House.

Vallat, R., and P.R. Ruby. 2019. "Is It a Good Idea to Cultivate Lucid Dreaming?" *Frontiers in Psychology* 10. *https://doi.org/10.3389/fpsyg.2019.02585*

Chapter 5

Bennett, C. 2018. *Liber 420: Cannabis, Magickal Herbs and the Occult*. Keneh Press.

Britten, E. H. 1876. *Art Magic; or Mundane, Sub-Mundane and Super-Mundane Spiritism*. Self-published.

Butler, W.E. 1991. *Magic and the Magician: Training and Work in Ritual, Power and Purpose*. Compilation edition. Aquarian Press.

Carroll, P. 1987. *Liber Null & Psychonaut*. Red Wheel/Weiser.

Eliade, M. 1964. *Shamanism: Archaic Techniques of Ecstasy*. Translated by Willard R. Trask. Princeton University Press.

Ellis, H. 1898. "Mescal: A New Artificial Paradise." *The Contemporary Review* 73. *www.samorini.it*

Fortune, D. 2000. *The Training and Work of an Initiate*. Rev. ed. Weiser Books.

Jay, M. 2019. *Mescaline: A Global History of the First Psychedelic*. Yale University Press.

Jay, M. 2023. *Psychonauts: Drugs and the Making of the Modern Mind*. Yale University Press.

Jünger, E. 1970. *Approaches: Drugs and Ecstatic Intoxication*. Available at: *files.libcom.org/files*

Karr, D., and S. Skinner, eds. 2010. *Sepher Raziel: Also Known as Liber Salomonis: A 1564 English Grimoire from Sloane MS 3826*. Golden Hoard Press.

Kingsland, J. 2019. *Am I Dreaming? The Science of Altered States, from Psychedelics to Virtual Reality, and Beyond*. Atlantic Books.

Knight, G. 1978. *A Practical Guide to Qabalistic Symbolism*. One-volume ed. Red Wheel/Weiser.

Lincoln Order of Neuromancers (L.O.O.N.). 1986. *Apikorsus: An Essay on the Diverse Practices of Chaos Magick*. Available at: *cdn.preterhuman.net*

Luke, D.P., and M. Kittenis. 2005. "A Preliminary Survey of Paranormal Experiences with Psychoactive Drugs." *Journal of Parapsychology* 69(2): 305–327. *researchgate.net*

Newman, P.D. 2011. "Magic Mirror Gazing Among the Rosicrucians." *Ad Lucem* XVIII.

Newman, P.D. 2017. *Alchemically Stoned: The Psychedelic Secret of Freemasonry*. The Laudable Pursuit Press.

Pagani, P. 1985. *Cardinal Rites of Chaos*. Sut Anubis.

Servants of the Light. n.d. "FAQ." Retrieved October 11, 2024, from *servantsofthelight.org*

Sheldrake, M. 2020. *Entangled Life: How Fungi Make Our Worlds*. Random House.

Stafford, P.G., and B.H. Golightly. 1967. *LSD—The Problem-Solving Psychedelic*. Award Books.

Valentino, T. 2020. "MDMA-Assisted Psychotherapy for PTSD Trials Progressing Through Phase 3." Psych Congress Network, 9 September. *hmpgloballearningnetwork.com*

Vayne, J. 2017. *Getting Higher: The Manual of Psychedelic Ceremony*. Psychedelic Press.

Wasson, R., A. Hofmann, and C. Ruck. 1978. *The Road to Eleusis: Unveiling the Secret of the Mysteries*. Harcourt Brace Jovanovich.

Whitby, C. 1988. *John Dee's Actions with Spirits: 22 December 1581 to 23 May 1583 (Volumes 1 and 2)*. Routledge.

Chapter 6

Eccles, J.C. 1989. *Evolution of the Brain: Creation of the Self*. Routledge.

Jung, C.G. 1991. *The Archetypes and the Collective Unconscious*. 2nd ed. Translated by R.F.C. Hull. Routledge.

Kaeuper, R.W., and E. Kennedy. 1996. *The Book of Chivalry of Geoffroi de Charny: Text, Context, and Translation*. University of Pennsylvania Press.

Oxford Languages. n.d. "Spirit." Retrieved October 11, 2024, from *google.com/search?q=spirit+definition*

Path of Joy. 2022. "Rumi Never Said: 'What You Seek Is Seeking You'!!" *youtu.be*

Chapter 8

Auryn, M. 2020. *Psychic Witch: A Metaphysical Guide to Meditation, Magick & Manifestation*. Llewellyn Publications.

Carroll, P. 1987. *Liber Null & Psychonaut*. Red Wheel/Weiser.

Everett III, H. 1957. "Relative State Formulation of Quantum Mechanics." Doctoral dissertation, Princeton University.

Green, M., and J.H. Schwarz. 1984. "Anomaly Cancellation in Supersymmetric D=10 Gauge Theory and Superstring Theory." *Physics Letters B* 149(2–3), 117–122.

Hernández, S. 2022. "Anil Seth: 'Reality Is a Controlled Hallucination.'" CCCB LAB, 22 November. *lab.cccb.org*

Hine, P. 1995. *Condensed Chaos: An Introduction to Chaos Magic*. New Falcon Publications.

Jung, C.G. 1991. *The Archetypes and the Collective Unconscious*. 2nd ed. Translated by R.F.C. Hull. Routledge.

Chapter 9

Chaotopia. 2023. "9 KP1." March 10, 3:26. *youtu.be*

Chapman, A. 2008. *Advanced Magick for Beginners*. Aeon Books.

Dixon, S.J. 2023. "Mastodon: Number of Registered Users 2022-2023." Statista, April 17. *statista.com/statistics*

IOT BIS. 2020. "The Illuminates of Thanateros - Evocation of Kawa Pohr." August 1, 11:23. *youtube.com*

Lee, D. 2008. *Bright from the Well: Northern Tales in the Modern World*. Mandrake of Oxford.

Lee, D. 2017. *Life Force: Sensed Energy in Breathwork, Psychedelia and Chaos Magic*. The Universe Machine.

Lee, D. 2023. *Primordial Chaos: Writings and Rituals from Then and Now*. The Universe Machine.

Lee, D. 2024a. "Egregore Against Weapons of Mass Destruction." *Chaotopia* (blog), January 15. *chaotopia-dave.blogspot.com*

Lee, D. 2024b. "A Chaotic Online Entity." *Chaotopia* (blog), January 30. *chaotopia-dave.blogspot.com*

Lee, D., and P. Mastin. 2020. "Kawa Pohr: The IOT's Healing Servitor." Illuminates of Thanateros - British Isles Section. *iotbritishisles.com*

Leslie, E. n.d. "Walter Benjamin's *Arcades Project.*" *Militant Esthetics* [blog]. Retrieved October 11, 2024, from *militantesthetix.co.uk*

Meijer, D.K.F. 2024. "On the Internet Meme/Virus Analogy: Part 1. Can We Prevent Contagious Information That Infects Our Sub-Conscious? A Plea for a Versatile Immune System for the Internet in the Present AI-Era." *academia.edu*

Chapter 11

Alli, A. 1985. *Angel Tech: A Modern Shaman's Guide to Reality Selection.* Vigilantero Press.

Alli, A. 2009. *The Eight-Circuit Brain: Navigational Strategies for the Energetic Body.* Vertical Pool Publishing.

Carroll, P. 1992. *Liber Kaos.* Red Wheel/Weiser.

Leary, J., and T. Leary 1973. *NeuroLogic.* Level Press.

Leary, T. 1977. *Exo-Psychology.* Starseed/Peace Press.

Wilson, R.A., and R. Shea. 1975. *The Illuminatus! Trilogy.* Dell.

Wilson, R.A. 1983. *Prometheus Rising.* New Falcon.

Chapter 12

Abrahamsson, C., ed. 2021. *The Mega Golem: A Womanual for All Times and Spaces.* Trapart Books.

Bradbury, R. 1951. *The Illustrated Man.* Doubleday & Co.

Dukes, R. 1999. *What I Did in My Holidays: Essays on Black Magic, Satanism, Devil Worship and Other Niceties.* Mandrake of Oxford.

Moore, S. 1995. Interview with Alan Moore. *Chaos International* 18.

Wicca, A. 1994. Interview with Frater Stokastikos. *Chaos International* 17.

Williams, R. 1994. Interview with William Burroughs. *Chaos International* 16.

Chapter 15

Carroll, P. 1992. *Liber Kaos.* Red Wheel/Weiser.

Carroll, P. 2014. *Epoch. The Esotericon & Portals of Chaos.* Arcanorium College.

Carroll, P. 2022. *Liber Null & Psychonaut.* Rev. ed. Red Wheel/Weiser.

Csikszentmihalyi, M. 1990. *Flow: The Psychology of Optimal Experience.* Harper Perennial.

Gómez Dávila, N. 2020. *O ľudskom údele*. Sol Noctis.

Greene, R. 2018. *The Laws of Human Nature*. Profile Books.

Evola, J. 1996. *The Doctrine of Awakening*. Inner Traditions.

Harris, R. 2011. *The Reality Slap: How to Survive and Thrive When Life Hits Hard*. Robinson.

Joseph, S. 2011. *What Doesn't Kill Us: The New Psychology of Postraumatic Growth*. Piatkus.

Kaufmann, W. 2013. *Nietzsche: Philosopher, Psychologist, Antichrist*. Princeton University Press.

Nietzsche, F. 1982. *The Portable Nietzsche*. Edited and translated by Walter Kaufmann. Penguin.

Reginster, B. 2006. *The Affirmation of Life*. Harvard University Press.

Taleb, N. 2012. *Antifragile: Things That Gain from Disorder*. Random House.

ENDNOTES

Chapter 3

1 For many examples, see Chapter 7, "Naive Intervention," in Nassim Taleb's *Antifragile*.

2 Make no mistake about the nature of Baphomet. He/She, as a progeny of Shub-Niggurath, is one of the Elder Gods—or more precisely, Outer Gods or Great Old Ones. Still young, weaker, and lesser compared to His/Her progenitors, but steadily learning and growing . . . at this stage of His/Her life cycle, mostly on us and through us, the billions of shoggoths, which He/She has created for this purpose.

3 Hippocrates's *era-primum non nocere* (first do no harm—avoid iatrogenic effects) should therefore also be a first principle of the wizard, in his own rational (multi-)self-interest.

4 The wand *and* the cup.

5 We see a similar tendency with Crowley's disciples. Those who took something from him and then cut off all contact with him had a chance to become quite successful—Major General J.F.C. Fuller, for example, or A.O. Spare. By contrast, those who stayed with him and continually submitted to his "initiatory syllabus" were usually ruined or at least debased.

6 From my anecdotal observation of zealous practitioners of dream yoga, as well as vipassana and zazen, it's not good. But it remains unclear what is cause and what is effect here. In Nietzsche's language, the very choice of these techniques and spiritual currents and their eager practice reveals decadence; their very choice is its consequence. In fact, we don't seem to choose our preferred doctrines and philosophies

freely. We get these decisions ready-made from the body. In a state of weakness or sickness, one tends to gravitate toward completely different ideologies than in a state of vitality, health, and high creativity. Our body tells us what it needs, depending on whether the will to life/power or the will to death/nothingness prevails in it. Secondarily, we "choose" the ideology that best resonates with this deep choice of the body and leads to appropriate behaviors/techniques. Both impulses are at work in each of us; in some phases of life the former may predominate, in another the latter. However, the overall tune and focus of a particular life is usually strongly affected by one or the other. Will to life/power: Chaos Magic, LaVeyan Satanism, Nietzsche, etc. Will to death/nothingness: Hermeticism, Buddhism, Christianity, Eastern religions, Hermetic magic, Western mystery tradition, Schopenhauer, etc. Thelema remains undecided. At the core it should belong to the first group, but in practice and theory it seems fatally contaminated to the FUBAR stage by elements from the second group.

7 John Yates, Daniel M. Ingram, Leigh Brasington, and other meditators acknowledge in their books that high doses of meditation are attempts to radically change/adjust/reprogram brain architecture and functioning through neuroplasticity. The risk/reward ratio of these attempts to "improve" the complex and unknown systems of the human brain is, to put it mildly, not so favorable.

8 For example, techniques of acceptance and commitment therapy.

9 If you still feel a compelling need to do something with yourself, always subtract first (your fragilities, harmful habits, and behavioral patterns). Don't add anything new, don't increase complexity. *Via negativa* as a first choice.

10 Liber MMM in Peter Carroll's *Liber Null & Psychonaut* provides pretty much all that you need in this regard. Perhaps also *A Liberated Mind* by Steven C. Hayes, *Flow* by Mihaly Csikszentmihalyi, and *The Six Pillars of Self-Esteem* by Nathaniel Branden. That's it, keep it simple.

11 This dynamic aim/state in Chaos Magic expresses the symbol of the eight-pointed star—the Octaris (as one of the variant meanings). Nietzsche said, "One must still have chaos in oneself to be able to give birth to a dancing star." On the notion of the "organizing" chaos of the inner bundle of drives as the ideal of Nietzsche's philosophy of power, see Chapter 9, "Power Versus Pleasure," in Walter Kaufmann's *Nietzsche: Philosopher, Psychologist, Antichrist*.

12 The best guide to this can be found in Mihaly Csikszentmihalyi's books *Flow* and *Creativity*. Also see Chapter 2, "Transform Self-Love into Empathy," in Robert Greene's *The Laws of Human Nature*.

13 In Pete Carroll's (2022) words: "Magic provides a high risk life strategy, the majority who adopt it seem to screw up, particularly the Chaos Magicians. In shortening the path I made it a lot more dangerous, but I offer no apologies for that."

14 Because for the vast majority of interested people, magic is an escape, a retreat from reality into illusion. Not an act of aggression (in the meaning of a movement toward reality and thus toward relationship to persons, things, situations, and experiences), but of regression (a withdrawal from the world, a retreat from reality, a fantasy).

15 You'll ignore it anyway. I know because as a young wand I also ignored all the well-meaning cautions. I made all the mistakes I'm about to warn against. Not all of my colleagues had as much dumb luck as I did to be able to report about it. In the midst of a magical fuck-up, this information will come in handy. Better late than never. You're welcome.

16 Thus in paganism it was believed that too much attention to one deity can provoke the anger and jealousy of others.

17 See the idea of the psychological shadow in Jungian cult language. According to this hypothesis, too much creation without corresponding destruction will result in the violent discharge of the destructive impulse, and so on. It's just a hypothesis, a bit crazy, but worthy of some attention. A useful book is Robert Johnson's *Owning Your Own Shadow: Understanding the Dark Side of the Psyche*.See also the idea of interconnection of the contents of the lower unconscious (negative) with the contents of the higher unconscious (positive) and their interplay in Roberto Assagioli's *Psychosynthesis*.Even in Freud's later teachings, the psyche is seen as a battlefield of two main interconnected impulses—the will to life and creation (Libido/Eros) and the will to death and destruction (Destrudo/Thanatos). Each of them has its own reservoir of psychic energy. Therefore, you should not ignore the wisdom of Thanateros and stick to the formula from *Principia Chaotica*: "Create, destroy, enjoy, IO CHAOS!" (Carroll 1992, 79).

18 Helpful books include Robert Johnson's *Inner Work—Using Dreams and Active Imagination for Personal Growth* and Eugene T. Gendlin's *Let Your Body Interpret Your Dreams*.

19 For example, astonishingly stupid impulsive action, unwanted temporary obsession, psychosomatic manifestation or illness, accidents, synchronicities, and the like. Some considerably damaging to the point of fatal if you induced too much imbalance too abruptly—by working with atavisms, nasty demons, or by persistently focusing on certain god-forms of your personal pantheon and ignoring others.

20 For a detailed explanation of these effects, see "Book V: The Nonlinear and the Nonlinear" in Taleb's *Antifragile.*

21 The best guidance for considering can be found in Chapter 8 of his book *The Octavo*, in the section "The Choice of What to Use Magic For."

22 More thoroughly: "We should not take risks with near-healthy people; but we should take a lot, a lot more risks with those deemed in danger. Why do we need to focus treatment on more serious cases, not marginal ones? Take this example showing nonlinearity (convexity). When hypertension is mild, say marginally higher than the zone accepted as 'normotensive,' the chance of benefiting from a certain drug is close to 5.6 percent (only one person in eighteen benefit from the treatment). But when blood pressure is considered to be in the 'high' or 'severe' range, the chances of benefiting are now 26 and 72 percent, respectively (that is, one person in four and two persons out of three will benefit from the treatment). So the treatment benefits are convex to condition (the benefits rise disproportionally, in an accelerated manner). But consider that the iatrogenics should be constant for all categories! In the very ill condition, the benefits are large relative to iatrogenics; in the borderline one, they are small. This means that we need to focus on high-symptom conditions and ignore, I mean really ignore, other situations in which the patient is not very ill" (Taleb 2012, 340–341).

23 See the essay "Magical Ritual" in Peter Carroll's book *The Octavo* for an explanation of what exactly these terms mean.

24 For example, sudden bursts of enchantments after a period of magical inactivity can be effective.

Chapter 4

1 The *Thieves' World* anthology, edited by Robert Asprin and Lynn Abbey.

Chapter 5

1 I use the term *bodymind* here to emphasize the notion of the body and mind as one thing.

2 One could argue that Michael Harner's "core shamanism," developed in the 1980s, represents a similar project to Chaos Magic, in that it seeks to identify and deploy common underlying techniques from multiple indigenous traditions.

3 The influence of Crowley on the formation of Chaos Magic cannot be overstated, especially in his attempt to create a psychological "scientific" magic, famously arguing for an occultist that would have "the method of science; the aim of religion." Following a life of strange drugs, black magic, and enthusiastic sex, Crowley died tragically young, at the age of 72.

4 Caxton Hall was known for hosting radical events, having been the site of the "Women's Parliament" started in 1907 by the Women's Social and Political Union, part of the British Suffragette movement.

5 Crowley's ritual was one of the first peyote ceremonies to take place outside of the Americas. For more detail, consult Mike Jay's *Mescaline: A Global History of the First Psychedelic.*

6 Classic psychedelics are a group of psychoactive drugs that including lysergic acid diethylamide (LSD), psilocybin, N,N-dimethyltryptamine (DMT), and mescaline (the principle active alkaloid in peyote). When ingested or administered, these compounds can elicit altered states of perception, cognition, and emotion, primarily by acting as 5-HT_{2A} (serotonin 5-hydroxytryptamine 2A) agonists in the brain.

7 The use of solanaceous plants, such as belladonna and henbane, is common in Western materia magica, but these are indeed very dangerous substances apt to cause, amongst other things, temporary blindness in the unwary. Cannabis, also commonly recommended in Western occult writing, is not in the same class of plants and is much less dangerous. While not a "classic psychedelic," it can certainly be said to have psychedelic activity, particularly when ingested.

8 British philosopher A.N. Whitehead once commented that the European philosophical tradition "consists of a series of footnotes to Plato." In much the same way, most Euro American trip reports could be characterized as footnotes to Ellis.

9 Research on using LSD for problem solving was conducted by the suitably named Eleusis Benefit Corporation in 2016. One participant, the biologist Merlin Sheldrake, narrates his attempt at using the chemognostic state to problem solve in his book *Entangled Life: How Fungi Make Our Worlds.*

Chapter 6

1 The mind-body problem refers to the unavoidable distinction between neurological and psychological activities. In fact, Sir John Carew Eccles, a neuroscientist who won the Nobel Prize for his brain studies, firmly concludes, "I maintain that the human mystery is incredibly demeaned by scientific reductionism, with its claim in promissory materialism to account eventually for all of the spiritual world in terms of patterns of neuronal activity. This belief must be classed as a superstition. . . . We have to recognize that we are spiritual beings with souls existing in a spiritual world as well as material beings with bodies and brains existing in a material world" (Eccles 1989, 24).

2 This quote is from Arishto's translation (Path of Joy, 2022).

Chapter 7

1 I use *Others* as a placeholder word for all apparently conscious or sentient entities. The whole category of spirits, be they land spirits, elementals, the dead, angels, demons, or gods.

2 I should add that I exclude quite a lot of artists and writers from this category, as some are anchored in very potent mythic reality. Primary examples of this are Alan Moore, Neil Gaiman, and Grant Morrison.

3 Chaos Magic often speaks of different "models" of magic. These include the spirit, energy, psychological, information, and meta models. These are put forward as ways you can structure (or analyze) magical workings and techniques.

4 The Field is my term for the whole of all that is and all that may be, the sum total of manifest and unmanifest reality.

5 While it is possible that there will be friction between you and the Others (or archetypes, if that's how you are perceiving them), part of how I structure the work is to minimize this.

Chapter 9

1 See Chaotopia (2023) for the Kawa Pohr pathworking with original sound, and IOT BIS (2020) for a version with more elaborate sound. For a discussion of Kawa Pohr history, see Lee and Mastin (2020).

2 For more information on the WABRI working, check out my blog post "Egregore Against Weapons of Mass Destruction" (Lee 2024a). You are invited to take part in this ongoing magical working and magical memetics experiment!

3 You can find a selection of some bits of that history in my blog post on John-Joan Mastodon (Lee 2024b).

4 A word of warning about how you choose to spend magical effort on spells against large corporations: in the mid-2010s I was present at two spellcastings against the irresponsible biotech giant Monsanto. Over the next year, we saw its share prices plummet. We felt pretty good until it was bought by Bayer, another gigantic company. The longer-term result had been not the loss of one destructive corporation, but the uniting of two of them. The takeaway lesson for me was: take more care in choosing precisely what you want to happen, and find the best point of ingress.

Chapter 10

1 Published under the pen name Ramsey Dukes.

2 Available at *https://www.youtube.com/@RamseyDukes/videos*.

Chapter 11

1 A phrase coined by Korzybski.

Chapter 14

1 Published in 1908 by the Yogi Publication Society and attributed to "Three Initiates" (widely believed to be the occultist and New Thought pioneer William Walker Atkinson).

2 The question was raised in a live Reddit Q&A on March 7, 2014. A transcript is available at *amatranscripts.com*

Chapter 15

1 For an elaboration of these insights, see the chapter on Baphomet in Peter J. Carroll's *Epoch: The Esotericon & Portals of Chaos*.

2 "In this new paradigm, the animating force of the entire vast universe can be called chaos. It is the inexpressible pregnant void from which manifest existence, order, and form arise. Being omnipresent and non-dualistic, it is virtually imperceptible, inconceivable, and impossible to visualize. Almost any attempt to say anything about it would be a denial of its other qualities and so a lie" (Carroll 2022, 139). Every living being carries within itself a "spark" of Will, which in Chaos Magic we call Kia (more precisely and less poetically: every living being functions as an "objectified" Kia). For this reason, Carroll correctly observed: "Kia cannot be experienced directly because it is the basis of consciousness (or experience), and it has no fixed qualities which the mind can latch on to" (Carroll 2022, 18).

3 "No image can fully represent the totality of what this force is, but it is conventionally shown as a hermaphroditic god-goddess in the form of a horned human that includes various mammalian and reptilian characteristics. It should also resume protozoan, insectivorous, and floral symbolism, for it is the animating spirit of everything from a bacterium to a tiger. . . . Between its horns a torch is usually positioned, for spirit is most easily visualized as light. The image should also include necrotic elements, for it also encompasses death. . . . The sexual aspects of the god-goddess Baphomet are always emphasized, for sex creates life, and the sexuality is a measure of the life force or vitality, no matter how it is expressed" (Carroll 2022, 140–141).

4 Perhaps because I consider the forest as the most potent death/birth generator, where this cycle is exhibited in great density and at high frequency. The forest, with its trees and groves, also corresponds best to how a human appears beneath the surface, beneath the illusion of a unified and orderly personality—just a chaotic mess of fractal processes clumping into clusters, all interconnected and at the same time competing with each other for resources and the satisfaction of their impulses. That the forest seems the best reflection of the human unconscious is not surprising, since both are products of the same fractal and blind natural creativity: Baphomet. (For both of these areas,

as for many other Baphometian domains, usually the less intervention in their natural attunement, the better—at least in the vast majority of cases, except for dangerous crises.)

5 In the fields of biology and medicine hormesis is defined as an adaptive response of cells and organisms to a moderate (usually intermittent) stress. Examples include exercise, dietary energy restriction, and exposures to low doses of certain phytochemicals. In other words, the individual organism is exposed to moderate doses of stress plus time for recovery and thus becomes stronger.

6 For a more detailed analysis, see Chapter 4, "What Kills Me Makes Others Stronger," in Nassim Taleb's *Antifragile*. Here is a quote to better illustrate: "Let us look at how evolution benefits from randomness and volatility (in some dose, of course). The more noise and disturbances in the system, up to a point, barring those extreme shocks that lead to extinction of a species, the more the effect of the reproduction of the fittest and that of random mutations will play a role in defining the properties of the next generation. Say an organism produces ten offspring. If the environment is perfectly stable, all ten will be able to reproduce. But if there is instability, pushing aside five of these descendants (likely to be on average weaker than their surviving siblings), then those that evolution considers (on balance) the better ones will reproduce, making the gene undergo some fitness. Likewise, if there is variability among the offspring, thanks to occasional random spontaneous mutation, a sort of copying mistake in the genetic code, then the best should reproduce, increasing the fitness of the species. So evolution benefits from randomness by two different routes: randomness in the mutations, and randomness in the environment—both act in a similar way to cause changes in the traits of the surviving next generations. Even when there is extinction of an entire species after some extreme event, no big deal, it is part of the game. This is still evolution at work, as those species that survive are fittest and take over from the lost dinosaurs—evolution is not about a species, but at the service of the whole of nature" (Taleb 2012, 68–69).

7 To get a better idea of your true relationship, consider that She/He doesn't even care about the whole species. Every extinction event, except total, is a field day for Her/Him.

8 However, according to Nietzsche, we should avoid a naively evolutionist view: "*My general view.—First proposition:* man as a species is not progressing. Higher types are indeed attained, but they do not last. The level of the species is *not* raised. *Second proposition:* man as a species does not represent any progress compared with any other animal. The whole animal and vegetable kingdom does not evolve from the lower to the higher—but all at the same time, in utter disorder, over and against each other" (Kaufmann 2013, 328).

9 "Everything in the universe establishes empires, and every single being wants to expand into all of creation. If even the most miserable animal were to give itself over to its fever, it would want to occupy the whole universe and devour the stars. In the fleeting organisms in the puddles of the pathways lies a hidden potential that wants to take possession of the heavens" (Gómez Dávila 2020, 6).

10 Baphomet follows the same principle on a larger level of scale and variety.

11 Some of the misconceptions of the will to power, including Nazi appropriation of Nietzsche's philosophy, arise from overlooking Nietzsche's distinction between *Kraft* (force) and *Macht* (power). *Kraft* is primordial strength that may be exercised by anything possessing it, while *Macht* is, within Nietzsche's philosophy, closely tied to sublimation and "self-overcoming," the intentional channeling of *Kraft* for creative purposes—mastery of something (skill, art, ability) and self-mastery.

12 Walter Kaufmann summed it up in a formula: power + joy, where joy includes not just elements of pleasure but also of pain. A detailed study of this formula can be found in the book *Flow* by Mihaly Csikszentmihalyi. An endless cycle of desire, overcoming resistance, and victory expressed in a runic formula for magical use: *jera*, *kenaz*, *nauthiz*, *sowilo*, *jera*.

13 Note that the will to power does not mean plain willpower, but seems intertwined with imagination. A deep desire manifests as imagination, the ability to envision, to see a desired goal or state, which acts as a motive in achieving.

14 I am by no means attempting here to make a complete map of human drives or to synthesize them. This eightfold division presents just a fairly rough model and symbolic illustration of how to work with your

drives and desires (and the corresponding memeplexes), which are vast in number and in constant flux.

15 Choose which term suits you. Some magicians prefer the playfulness of games, others the seriousness of quests or the non-poetic power of organic drives. Others alternate between these approaches.

16 Not to achieve and stop, but to continually achieve until the death of the magician.

17 The branch of positive psychology has proved this quite convincingly. See books by Mihaly Csikszentmihalyi and Martin Seligman.

18 I am using the traditional astrological designation for the basic division, but in the Chaos Magical sense as applied by Pete Carroll in *Liber Kaos* and *Epoch*.

19 In fact, they represent the eight main branches of your tree of drives and memeplexes, each of which sprouts other subsidiary drives and desires. The psyche resembles a tree, the roots and trunk representing the basal principle of the will to power, the main branches the eight major drives with many minor branches, etc. This whole environment has a fractal structure, so in the same way we can say that the psyche resembles a wood, which breaks up into eight large groves, and so on. (You can design your temple for astral magic conjurations in this way, in the shape of Octaris. See Liber KKK in *Liber Kaos*.)

20 Determined by genetics, specific wiring of the brain, and character/sequence of experiences in the early years of life. See Chapter 13, "Advance with the Sense of Purpose," in *The Laws of Human Nature* by Robert Greene.

21 For instructions on conjurations for each of the eight types of magic, see Chapter 2, "Sleight of Mind," and Chapter 4, "Eight Magics," in Pete Carroll's *Liber Kaos*.

22 For example, as part of my mercurial quest/game, I have written books, television and movie screenplays, commercials, political speeches, news articles, marketing campaigns, TV reportages, documentaries, and texts for museum exhibitions—all different mercurial sub-games (or even sub-sub-games within the mercurial sub-game "writing"). I suppose if I live long enough, the list will get broader. The situation looks similar in my other major games.

23 Since these will inevitably strike (you just don't know in which games, in what form or sequence or when), you should anticipate them and

prepare in advance by learning the skills to cope with them. See Russ Harris's *The Reality Slap* and Stephen Joseph's *What Doesn't Kill Us* for more on this.

24 "The goal in itself is usually not important; what matters is that it focuses a person's attention and involves it in an achievable, enjoyable activity. . . . The unrelated goals of the separate flow activities merge into an all-encompassing set of challenges that gives purpose to everything a person does. . . . It is not enough to find a purpose that unifies one's goals; one must also carry through and meet its challenges. The purpose must result in strivings; intent has to be translated into action. We may call this resolution in the pursuit of one's goals. What counts is not so much whether a person actually achieves what she has set out to do; rather, it matters whether effort has been expended to reach the goal, instead of being diffused or wasted" (Csikszentmihalyi 1990, 216–217).

TO OUR READERS

Weiser Books, an imprint of Red Wheel/Weiser, publishes books across the entire spectrum of occult, esoteric, speculative, and New Age subjects. Our mission is to publish quality books that will make a difference in people's lives without advocating any one particular path or field of study. We value the integrity, originality, and depth of knowledge of our authors.

Our readers are our most important resource, and we appreciate your input, suggestions, and ideas about what you would like to see published.

Visit our website at *www.redwheelweiser.com*, where you can learn about our upcoming books and free downloads, and also find links to sign up for our newsletter and exclusive offers.

You can also contact us at *info@rwwbooks.com* or at

Red Wheel/Weiser, LLC
65 Parker Street, Suite 7
Newburyport, MA 01950

TO OUR READERS